Sensing and Artificial Intelligence Solutions for Food Manufacturing

This book gives readers a practical introduction into machine learning and sensing techniques, their design and ultimately specific applications that could improve food production. It shows how these sensing and computing systems are suitable for process implementation in food factories.

This book starts by giving the reader an overview of the historic structures of food manufacturing standards and how they defined today's manufacturing. It is followed by a topical introduction for professionals in the food industries in topics such as AI, machine learning, and neural networks. It also includes an explanation of the different sensor systems and their basic principles. It shows how these sensing and computing systems are suitable for process implementation in food factories and what types of sensing systems have already been proven to deliver benefit to the food manufacturing industries. The authors also discuss issues around food safety, labelling, and traceability and how sensing and AI can help to resolve issues. They also use case studies and specific examples that can show the benefit of such technologies compared to current approaches.

This book is a practical introduction and handbook for students, food engineers, technologists and process engineers on the benefits and challenges around modern manufacturing systems following Industry 4.0 approaches.

Sensing and Artificial Intelligence Solutions for Food Manufacturing

Edited by

Daniel Hefft and Charles Oluwaseun Adetunji

CRC Press
Taylor & Francis Group
Boca Raton London New York

CRC Press is an imprint of the
Taylor & Francis Group, an **informa** business

Designed cover image: Shutterstock

First edition published 2023
by CRC Press
6000 Broken Sound Parkway NW, Suite 300, Boca Raton, FL 33487-2742

and by CRC Press
4 Park Square, Milton Park, Abingdon, Oxon, OX14 4RN

CRC Press is an imprint of Taylor & Francis Group, LLC

ISBN: 978-1-032-07580-8 (hbk)
ISBN: 978-1-032-07618-8 (pbk)
ISBN: 978-1-003-20795-5 (ebk)

DOI: 10.1201/9781003207955

Typeset in Times
by Newgen Publishing UK

Contents

The Editors

Daniel Hefft is a food engineer from Germany. He received his undergraduate degree from Technische Hochschule Ostwestfalen-Lippe with specialization in cereal technology. He has a master's degree from the University of Reading and is currently at the University of Birmingham. He has been awarded research fellowships and research affiliation with the University of Birmingham. He has extensive years of experience in the food industry, having worked for a range of companies. His specialties are food-process design and engineering. He is the founder and past CTO of Rheality Ltd., which utilizes a novel technology based on acoustic sensing and machine learning for rheology measurements. Daniel is also the founder of the Engineering for Food and Drinks Special Interest Group at the Institution of Agricultural Engineers and its current chair. Since 2018 he has gained professional recognition as a chartered engineer with the Institution of Agricultural Engineers. He is further involved in the Association of German Food Technologists (GDL e.V.) and the Institute of Physics (IOP).

Charles Oluwaseun Adetunji is presently a faculty member at the Microbiology Department, Faculty of Sciences, Edo State University Uzairue (EDSU), Edo State, Nigeria, where he utilized the application of biological techniques and microbial bioprocesses for the actualization of sustainable development goals and agrarian revolution, through quality teaching, research, and community development. He was formally the Acting Director of Intellectual Property and Technology Transfer, the Head of Department of Microbiology, Sub Dean for Faculty of Science and currently the Chairman Grant Committee and the Ag Dean for Faculty of Science, at EDSU. He is a Visiting Professor and the Executive Director for the Center of Biotechnology, Precious Cornerstone University, Ibadan, as well as a visiting scientist to the Department of Biotechnology and Food Science, Durban, South Africa. He is presently an external examiner to many academic institutions around the globe most especially for PhD and MSc students. He has won several scientific awards and grants from renowned academic bodies such as Council of Scientific and Industrial Research (CSIR), India; Department of Biotechnology (DBT), India; The World Academy of Science (TWAS), Italy; Netherlands Fellowship Programme (NPF), Netherlands; The Agency for International Development Cooperation, Israel; and Royal Academy of Engineering, UK. He has published many scientific journal articles and conference proceedings in refereed national and international journals with over 450 manuscripts and 4 scientific patents. He was ranked recently as number 20 among the top 500 prolific authors in Nigeria between 2019 till date by SciVal/SCOPUS. His research interests include Microbiology, Biotechnology, Post-harvest management, Food Science, Bioinformatics, and Nanotechnology. He was recently appointed as the President and Chairman Governing Council of the Nigerian Bioinformatics and Genomics Network Society. He was also recently appointed as the Director for International Affiliation and Training Centre for Environmental and Public Health, Research and Development, Zaria. He is presently a series editor with Wiley, Elsevier,

Taylor and Francis, USA, editing several textbooks on Agricultural Biotechnology, Nanotechnology, Pharmafoods, and Environmental Sciences. He is an editorial board member of many international journals and serves as a reviewer to many double-blind peer review journals published by Elsevier, Springer, Taylor and Francis, Wiley, PLOS One, Nature, American Chemistry Society, Bentham Science Publishers etc. He is a member of many scientific and professional bodies including American Society for Microbiology, Biotechnology Society of Nigeria, and Nigerian Society for Microbiology, and is presently the General/Executive Secretary of Nigerian Young Academy. He is the National coordinator for South-West Nigerian Society for Microbiology. He has won a lot of international recognition and also acted as a keynote speaker delivering invited talk/position paper at various universities, research institutes, and several centers of excellence that span across several continents of the globe. He has over the last fifteen years built strong working collaborations with reputable research groups in numerous and leading universities across the globe. He is the convener for Recent Advances in Biotechnology, which is an annual international conference where renown microbiologists and biotechnologists come together to share their latest discoveries. He is the president and founder of the Nigerian Post-Harvest and Food Biotechnology Society.

Contributors

Inobeme Abel – Department of Chemistry, Edo State University Uzairue, Nigeria

Ogundolie Frank Abimbola – Department of Biotechnology, Baze University Abuja, Nigeria

Charles Oluwaseun Adetunji – Biotechnology Research Group, Department of Microbiology, Edo State University Uzairue, Edo State, Nigeria

Juliana Bunmi Adetunji – Nutritional and Toxicological Research Laboratory, Department of Biochemistry Sciences, Osun State University, Osogbo, Nigeria

Alexander Ikechukwu Ajai – Department of Chemistry of the Federal University of Technology Minna, Nigeria

Modupe Doris Ajiboye – Plant Science and Biotechnology, Faculty of Science, Federal University, Oye Ekiti, Nigeria

Olalekan Akinbo – AUDA-NEPAD Centre of excellence in Science, Technology, and Innovation, Stellenbosch, South Africa

Osikemekha Anthony Anani – Laboratory of Ecotoxicology and Forensic Biology, Department of Biological Science, Faculty of Science, Edo State University Uzairue, Edo State, Nigeria

Tsegaye Bojago Dado – Department of Physics, College of Natural and Computational Sciences, Wolaita Sodo University, Ethiopia

Babatunde Joseph Dare – Department of Anatomy, Osun State University, Oshogbo, Nigeria

Wadazani Dauda – Department of Crop Protection, Federal University Gasua, Nigeria

Aruwa Christiana Elejo – Department of Biotechnology and Food Science, Durban University of Technology, Durban, South Africa

Shakira Ghazanfar – Functional Genomics and Bioinformatics, National Agricultural Research Centre, Islamabad, Pakistan

Onyijen Ojei Harrison – Department of Mathematical and Physical Sciences, Samuel Adegboyega University Ogwa, Edo State, Nigeria

Daniel Hefft – Edgbaston Campus, University of Birmingham, School of Chemical Engineering, Birmingham B15 2TT, United Kingdom

Inamuddin – Department of Chemistry, King Abdulaziz University, 21589 Jeddah, Saudi Arabia

Jonathan Inobeme – Department of Chemistry, Ibrahim Badamasi University Lapai, Niger State, Nigeria

Maliki Munirat – Department of Chemistry, Edo University Iyamho, Nigeria

Manjia Jacqueline Njikam – Laboratoire of Pharmacology and Toxicology, Department of Biochemistry, Faculty of Science, University of Yaoundé I, P.O. Box 812 Yaoundé, Cameroon

Michael O. Okpara – Department of Biochemistry, Federal University of Technology, Akure, Ondo State, Nigeria

Olugbemi T. Olaniyan – Laboratory for Reproductive Biology and Developmental Programming, Department of Physiology, Rhema University, Aba, Nigeria

Akinola Samson Olayinka – Computational Science Research Group, Department of Physics, Edo State University Uzairue, Edo State, Nigeria

Tosin Comfort Olayinka –Department of Information Technology and Cyber Security, College of Computing, Wellspring University, Benin City, Edo State, Nigeria

Olorunsola Adeyomoye – Department of Physiology, University of Medical Sciences, Ondo City, Nigeria

Osarenkhoe O. Osemwegie – Microbiology Department, Landmark University (SDG 12), Omu Aran, Kwara State, Nigeria

John Tsado Mathew – Department of Geography, Ahmadu Bello University Zaria, Nigeria

1 Overview of Sensors and Sensing Techniques in Food Processing

Akinola Samson Olayinka, Tsegaye Bojago Dado, Charles Oluwaseun Adetunji, Tosin Comfort Olayinka, and Onyijen Ojei Harrison

CONTENTS

1.1 INTRODUCTION

Sensors can detect and convert physical parameters to measurable electrical signals that can be processed for further action and/or decisions. Sensors can be used to detect several parameters like temperature, light, humidity, motion, or pressure from the environment where they are installed. The electrical signals from sensors can be processed by a microcircuit processor to produce datasets that can be sent to receivers for further processing and applied for various uses and decision making. Sensors come in different variants and functionalities and the specific domain of application influences the choice per time. Most operations in the production environment towards quality and efficiency require proper monitoring of the process leading to the production. To achieve this, sensor technology plays a significant role and the need to choose the right sensor for specific application cannot be over-emphasized. To

DOI: 10.1201/9781003207955-1

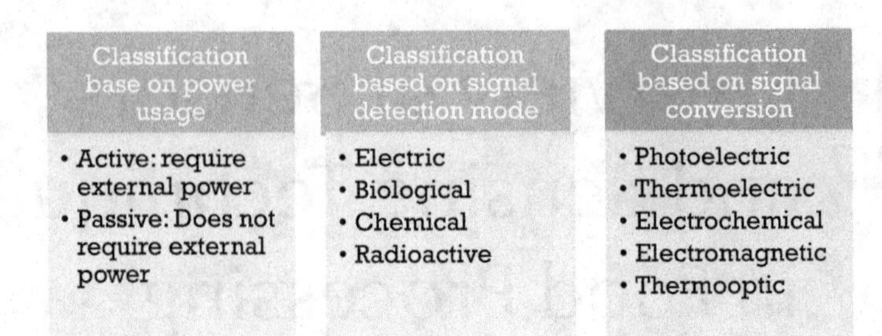

FIGURE 1.1 Sensor Classifications.

meet these requirements, sensors must be durable, flexible, and reliable in any environment. The capabilities, limitations, and applicability of a sensor must be carefully considered when choosing one. Sensors and other emerging technologies like artificial intelligence and machine learning are essential components of the food industry's monitoring and control system to avoid process failure (Misra et al. 2020; Olayinka et al. 2022; Kovalenko and Chuprina 2022). Sensors are based on the physical, chemical, or biological reaction of the environment. High-specificity detection is required for food and beverage components as well as food- and water-borne diseases, toxins, and pesticide residues. The magnitude of the real world is transformed into useable signals by sensors, which are then converted to an electric signal and sent to a display system that can be read (Steinberg et al. 2019). Sensor technologies have been applied in several food-related activities lately. Prominent among them are food safety and quality control against contaminations and preservation of the food chain (Curulli 2021; M. Lv et al. 2018; Balbinot et al. 2021; Hua et al. 2021; Kumar and Neelam 2016); food processing and monitoring (Thakur and Ragavan 2013; Villalonga et al. 2022); food borne pathogens identification and control (Yunus and Kuddus 2020; Wu et al. 2019); food, water and drug analysis (Kurbanoglu, Erkmen, and Uslu 2020; Li et al. 2019; Karimi-Maleh et al. 2020; Lin et al. 2021). Figure 1.1 shows various classifications of sensors based on power usage, detection mode and signal conversion method. In the choice of sensors for various applications, there are a number of factors that are essential for consideration before a particular sensor is adopted in any application. Figure 1.2 shows factors that should be considered when making such choices.

1.2 SENSORS IN FOOD INDUSTRY

Various sensors are used in the food processing industry for diverse applications due to their versatility and high level of functionality. Such sensors include proximity sensors (inductive, capacitive, and ultrasonic type), temperature sensor (resistance temperature detector, infrared sensor, thermistor, and thermocouple type), humidity sensors (optical, gravimetric, capacitive, resistive, piezoresistive, and magnetoelastic type), biosensors (amperometric, conductometer, thermometric biosensor, and potentiometric type), chemosensory sensors, pressure sensors, electronic tongue (E-tongue) taste sensors, torque sensors, freshness sensors, pH sensors and gas sensors.

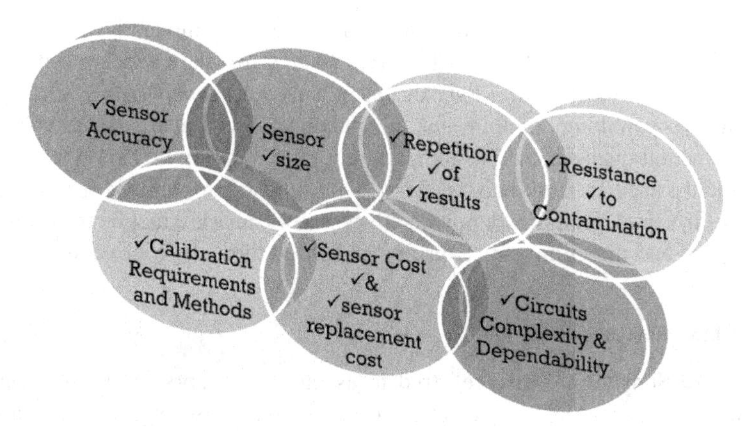

FIGURE 1.2 Factors Influencing the Choice of Sensors.

1.2.1 Proximity Sensors

Proximity sensors detect the presence of objects within a certain distance without requiring physical contact. A proximity sensor sends an electromagnetic radiation (EMR) such as infrared on a regular schedule and watches for changes in the return signal. Even when working at fast speeds, proximity sensors are intended to provide precise and reliable readings. These high-speed industrial sensors, which can operate at up to 5,000 Hz, are ideal for a variety of applications (Nguyen et al. 2021; Ye et al. 2020). The various types of proximity sensors are as follows:

1. Inductive sensors: Metal objects are detected by these (ferrous or nonferrous). There is a wider sensing range for ferrous metals, ranging from 5mm to 40mm. Non-ferrous metals do not contain iron, which can reduce detection by up to 60 percent. An electrical current known as an eddy current is induced when a ferrous material is introduced into a magnetic field on the surface. This eddy current causes a power loss on a metal surface when it passes through it. Because inductive sensors can function at higher rates than mechanical switches, they are commonly utilized. Inductive sensors are more reliable and less susceptible to failure. Sinambela *et al.* used inductive sensors in the monitoring of palm plantation for predicting harvest time for ripe palm fruit (Sinambela et al. 2020).
2. Capacitive sensors: Capacitive sensors can detect dielectric material (both higher and lower dielectric material) such as liquid, glass, plastic wood, and granulated substance. These capacitive sensors can detect dielectric constants as high as 1.2. Two plates are in the sensory form of capacitive sensors. The capacitance of the two plates increases as the target approaches the sensor range, causing the accelerated frequency to change. This makes it easier to identify the dielectric material in the food. It has been used in sachet seal and container capping integrity test in non-destructive evaluation of food products (Mohan, Toh, and Malcolm 2021b; 2021a).

3. Ultrasonic sensors: Ultrasonic sensors can work in smoke, dust, fog, different colors, textures, and steam. These ultrasonic sensors are widely used. This can detect an object from a long distance. This works well with solid materials such as gar, flour, potato, water, oil, and juice. This is also unaffected by environmental changes such as temperature, noise, and light. This sensor is commonly used in automated sensors. This sensor can identify objects that are solid, liquid, granular, or powdered, as well as measure factors including wind speed and direction, tank capacity, and speed through air or water.

1.2.2 Temperature Sensors

Temperature Sensors are also referred to as thermo-sensors. A thermo-sensor is a gadget that is capable of measuring the temperature of an object or an environment where it is located. A temperature sensor is one of the most prominent sensors in the Internet of Things (IoT) applications and wearable technologies. Recent developments have seen the use of flexible temperature sensors in various applications in food processing and other similar technologies (Su et al. 2020; Arman Kuzubasoglu and Kursun Bahadir 2020; Nwankwo, Olayinka, and Ukhurebor 2019). Process control, food inspection, freezers, fermentation units, baking ovens, and grills and smoking units are just a few of the uses of these thermosensors in the food industry. Some of the prominent types of thermos sensors are:

1. Negative temperature coefficient (NTC) thermistor: A thermally sensitive resistor is a thermistor that responds to temperature changes with a large, predictable, and exact change in resistance. The effective working temperature range of the NTC thermistor is -50 to 250 ° C.
2. Detector of temperature resistance (RTD): An RTD (resistance thermometer) is a type of thermometer that measures resistance. When the resistance of the RTD is compared to the temperature, the RTD can be used to determine the temperature. RTDs can be used at temperatures ranging from -200 to 600 degrees Celsius.
3. Thermocouple: A thermocouple usually consists of two wires that are connected in two places. The wires are metals that are of different elemental composition. The voltage difference between these two sites represents proportionate temperature variations. The precision of a thermocouple ranges from 0.5 to 5 degrees Celsius. They do, however, operate in the broadest temperature range, ranging from -200 to 1750 degrees Celsius.
4. The most sensitive temperature sensors are those made of semiconductors. In addition, they have the slowest reaction (5 to 60 seconds) in the narrowest temperature range (-70 to 150 degrees Celsius).
5. Infrared thermometers: This thermometer uses an electromagnetic wave infrared light to scan the object and the infrared detector processes the absorbed energy into an electrical signal, which is then interpreted to temperature. Infrared thermometers have been used in the measurement of food temperature in food-processing setting (Rohit et al. 2019).

1.2.3 Humidity Sensors

Humidity is defined as the quantity of water vapor in the air, which may be a combination of air and a pure gas like nitrogen or argon. It can have an impact on human comfort as well as on a wide range of production processes in sectors such as electronics, food and pharmaceutical manufacture, food storage, and so forth. The water vapor affects a variety of physical, chemical, and biological properties. Furthermore, correct humidity levels are vital to product quality and can help conserve energy. Humidity management is critical in the purification of chemical gases, ovens, dryers, paper and textile manufacturing, and food preparation. Humidity measurement is critical in powder manufacturing, especially when working with hygroscopic materials.

Humidity sensors are devices that detect and transmit the wetness and air temperature of a given environment, such as in the air, soil, or a restricted place, where such sensors are located. Humidity readings reveal the amount of water vapor present in the atmosphere. Capacitive, resistive, and thermal humidity sensors are the three most common kinds. To compute the humidity in the air, all three types will detect minute changes in the environment. Humidity sensors are a basic type of sensors yet they are crucial to various applications in our daily lives and industrial processes (Hussein et al. 2020; Nooriman et al. 2018; Shun et al. 2020; Liang et al. 2020; Kim et al. 2021; C. Lv et al. 2019). Different humidity sensors like graphene-based nano type in intelligent food packaging systems (Moustafa et al. 2021) and low cost type in IoT based food packaging applications (Popa et al. 2019). Humidity sensors include capacitive humidity sensor; resistive humidity sensor; thermal humidity sensor; gravimetric humidity sensor; optical humidity sensor; and piezoresistive humidity sensor.

1. Capacity Humidity Sensor: This is one of the most prevalent forms of humidity sensors, and it operates by the capacitive effect. They are frequently employed in situations where cost, stiffness, and size are all essential considerations. The electrical permittivity of the dielectric material in capacitive relative humidity (RH) sensors fluctuates as the humidity changes. A small amount of capacitance is formed between the pair of electrodes and the dielectric material. Plastic or polymers are the most common dielectric materials used in capacitive sensors, with dielectric constants ranging from 2 to 15. The capacitance value is calculated in the absence of moisture by determining the dielectric constant and the sensor geometry. The dielectric constant of water vapors is approximately 80 at room temperature, which is greater than the dielectric constant of the sensor. As a result, as water vapor is absorbed, the capacitance sensor values rise (Komazaki and Uemura 2019). These sensors are made up of an air-filled capacitor that changes its permittivity as the humidity in the air changes. Because air is a dielectric, it cannot be used for practical purposes. As a result, the gap between the capacitor plates is often filled with a suitable dielectric substance (isolator), the dielectric constant of which fluctuates with humidity. The most popular techniques to build a capacitive RH sensor are using a hygroscopic polymer sheet as the dielectric and two layers of electrodes on either side. Using this sensor, the change in frequency of the test sample is determined. The sample is inserted between the capacitor and the

oscillator is produced utilizing capacitance with RH sensitive test as dielectric throughout the procedure. Capacitive humidity sensors are used in a variety of applications such as weather stations, automobiles, food preparation, refrigerators, ovens, clothes dryers, printers, and high-voltage AC systems to absorb moisture from the system. Capacitive humidity sensors have benefits such as:

- The variation of the output is nearly linear in nature.
- They produce consistent results over time.
- They can be used in a wide range of relative humidity
- It allows for a shorter distance between the sensor and the signaling circuit.

2. Resistive Humidity Sensors (Electrical Conductivity Sensors): Resistive humidity sensors detect changes in the resistance value of the sensor element as humidity changes. The key assumption is that the conductivity of nonmetallic conductors is dictated by their water content. These sensors are comprised of a low-resistance material. The resistivity of the material fluctuates as the humidity changes. The connection between resistance and humidity in nature is inversely exponential. The low-resistance material is placed on top of the two electrodes, which are inter-digitized to increase the contact area. When the top layer absorbs water, the resistivity between the electrodes varies, which may be measured using a simple electric circuit. Modern sensors are coated with a ceramic material to provide further protection. Sensor electrodes are often made of precious metals such as gold, silver, or platinum. Resistive humidity sensors have the disadvantage of being particularly sensitive to chemical vapors and other pollutants. When water-soluble materials are used, the output measurements may vary.
3. Thermal Conductivity Humidity Sensors: Because these sensors detect absolute humidity, Absolute Humidity (AH) sensors are another name for them. This sensor detects the thermal conductivity of dry air as well as water-vapor-containing air. When developing a moisture sensor based on thermal conductivity, the ideal component to employ is a thermistor. Consequently, two small thermistors with negative temperatures were created with coefficients to form a bridge circuit. One thermistor is maintained open to the air via a small venting hole, while the other is hermetically sealed in a dry nitrogen chamber. The circuit is turned on, and the resistance of the two thermistors is measured. The difference in thermistor resistance is proportional to absolute humidity (AH). Benefits of thermal conductivity humidity sensors includes suitability for high corrosive and high temperature environments as well as higher resolution. Thermal Conductivity Humidity Sensors are used in drying pharmaceutical plants and other drying machines. They are used in food dehydration processes.
4. Optical fiber humidity sensors (OFHS) are used. They offer various advantages over electronic humidity sensors, including a tiny size, robustness, the ability to operate in combustible environments, as well as at higher temperatures

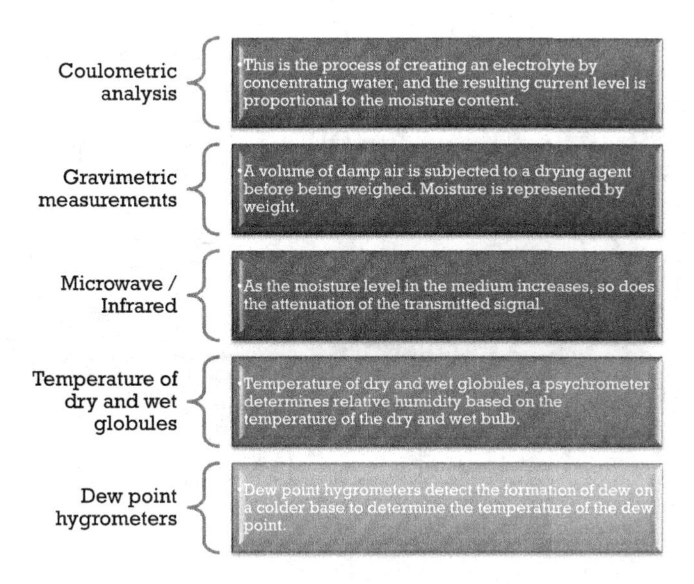

FIGURE 1.3 Mechanisms of Humidity Sensing.

and pressures and, most importantly, electromagnetic immunity. As a result, they can endure the harsh and demanding conditions inherent in processing methods.

Mechanisms of Humidity Sensing is shown in Figure 1.3.

1.2.4 Biosensors

Biosensors are devices that measure the concentration of an analyte. In biosensors, biological material (enzymes, antibodies, whole cells, and nucleic acids) is used to interact with the analyte. This contact causes a physical and chemical change, which the transducer detects and translates to an electric signal. The electric signal is proportional to the concentration of the analyte. This signal is decoded and transformed to the analyte concentration in the sample. Biosensors have been applied to pest and pathogens control in agricultural-related applications (Adetunji et al. 2021). Biosensors are usually constructed from three components: receptor (biological compound), transducer, and electronics. Figure 1.4 shows the block diagram of a typical biosensor.

The various types of biosensors are as follows:

1. Potentiometric biosensors: ISFETs (ion-sensitive field effect transistors) and ISEs (ion-sensitive electrochemical sensors) are the foundations of potentiometric biosensors (ion-selective electrodes). This sensor detects potential differences that occur during redox reactions (oxidation-reduction reaction). A redox reaction is a chemical reaction in which the number of oxidations of

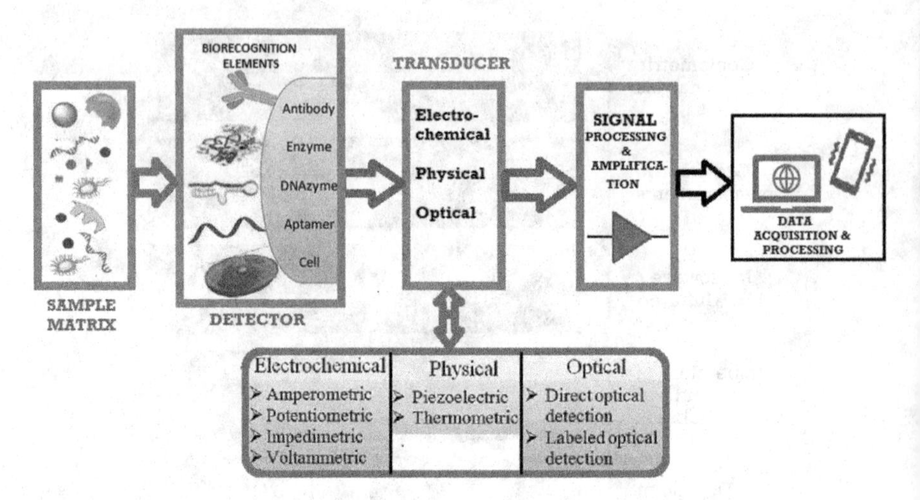

FIGURE 1.4 Biosensor Block Diagram.

a molecule, atom, or ion change by the acquisition or loss of an electron. Ions assembling at the ion-selective membrane interface might account for most of the output signal. The current passing through the electrode is either zero or close to zero. As a result of the enzyme reaction, the electrode detects the existence of the monitored ion (Mulchandani et al. 1999; Law Al and Adeloju 2013; Kauffmann and Guilbault 1991).

2. Amperometric biosensors: Theses are instruments that provide specific quantitative analytical information by measuring the current generated by the oxidation or reduction of an electroactive biological element. Amperometric biosensors are preferable to potentiometric biosensors because they have a higher sensitivity in comparison. (Ghindilis et al. 1998). A noble metal is commonly used in the biorecognition component of amperometric biosensors and has been recognized as a useful tool for monitoring drug administration in clinical trials and disease management (Wang 1999; Goud et al. 2019). Carbon paste with an integrated enzyme is another low-cost solution (Cui, Liu, and Lin 2005). Amperometric biosensors measure the flow of electrons that occurs during the reaction. Hayat et al. presented different perspective on the nano-based amperometric biosensors and its potential to provide an interesting development in biosensors and its applications (Hayat, Catanante, and Marty 2014). Zadeh et al. investigated how the geometry of perforated membrane influence the functionalities of biosensors. They studied the impacts of the porosity depth of the perforated membrane, the enzyme saturation level, and the half-time responsiveness on the biosensor's instantaneous and steady-state currents. Their simulation results concluded that the biosensor's instantaneous and steady-state currents are influenced by the geometry of perforated membrane, thereby affecting the biosensor's sensitivity (Zadeh et al. 2020).
3. Thermometric biosensors: Heat absorption (endothermic response) or release (exothermic reaction) is related to a variety of different biological

processes and is a major component of thermometric biosensors. Thermal process have use in various sensor applications such as clinical practices, monitoring and measurements, and oncoheamatology practices, pathogen detection and drug detection (Spink and Wadsö 1976; Naresh and Lee 2021; Fracchiolla, Artuso, and Cortelezzi 2013; Mehrotra 2016; Ramanathan and Danielsson 2001). Calorimetric biosensors are also known as thermometric biosensors. In thermometric biosensors changes in heat are directly proportional to the extent of reaction (for catalysis) or structural dynamics of biomolecules in the dissolved state. Colorimetric biosensors commonly used in the estimation of food chemical hazard, such as synthetic food additives, antibodies, residues of veterinary drugs, heavy metals, and other toxic substances.

4. Fiber optic lactate biosensor: It measures the change in oxygen concentration, and molecules change as a result of the oxygen effect in fluorescent dye. The enzyme lactate monooxygenate reduces the reaction below. This is commonly used in the detection of food pathogens.
5. Conductometric biosensors: During a reaction, this instrument measures changes in electrical conductivity. A common example is urea biosensors.
6. Optical biosensors: These devices measure the light emitted by the enzyme. Florescence, absorbance, and other principles are used in optical biosensors. To name a few, pressure measuring sensors are also called piezometers, pressure transmitters, pressure transducers, pressure indicators, pressure transmitters, and manometers.

1.2.5 Pressure Sensor

A pressure measurement sensor or instrument is a device that measures and converts pressure (a physical variable) into an electrical signal. Depending on the amount of pressure applied, the electrical signal may change. Pressure sensors can detect pressures that are lower than atmospheric pressure. A pressure sensor can also be used to measure other variables such as water level, water level, and altitude. Pressure measurement devices include absolute pressure sensors, gauge pressure sensors, vacuum pressure sensors, differential pressure sensors, and sealed pressure sensors. The vacuum sensor is one of the most widely used pressure-sensing devices. Vacuum sensors are commonly used to detect pressures lower than atmospheric pressure. The medium and high vacuum may be measured using thermal and molecular instruments. A vacuum sensor's detecting element is a heating element, the temperature of which is determined by the enclosed pressure. The temperature of the heating element increases as the sub-atmospheric pressure increases.

1.2.6 Electronic Noses

Electronic noses are instruments that can detect odors more effectively than the human nose. This device is more accurate and precise, and it detects odors using a chemical action mechanism. This can identify dangerous and noxious odors that the human nose cannot detect. These sensors are widely used in agriculture, the food

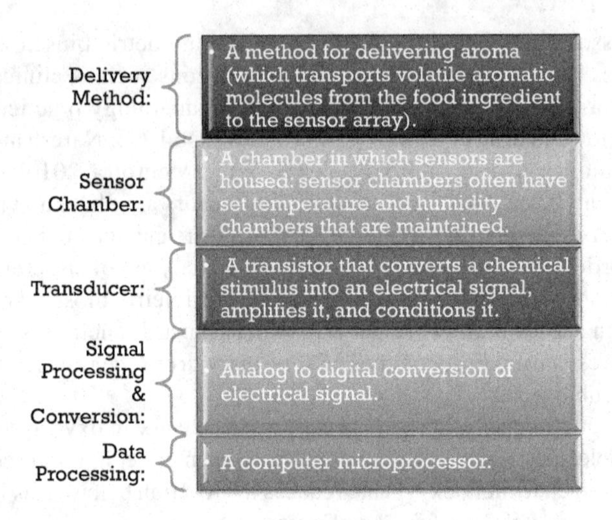

FIGURE 1.5 Mechanisms of the Electronic Nose Sensor System.

processing industry, public security, cosmetics, biochemistry, and other sectors of scientific research. The mechanism used by the electronic nose sensor device to identify aromatic compounds in food is shown in Figure 1.5.

1.2.7 Electronic Tongue

The electronic tongue (E – tongue) mimics the gustatory system of humans. It artificially reproduces the taste sensation. It is capable of detecting sourness, astringency, bitterness, saltiness, and umami. The e-tongue can be used in both qualitative and quantitative applications. This instrument includes a sensor array, electrochemical cells such as conductimetric, amperometric, potentiometric, impedimetric, and voltametric cells, as well as an appropriate pattern recognition system, which can easily recognize the sample as simple or complex nonvolatile soluble molecules (Kayumba et al. 2007; Emanuelli et al. 2021; Pradhan et al. 2020; Zheng and Keeney 2006; Sadrieh et al. 2005). These sensors are most suitable for applications in all food processing industries such as beer, wine, coffee, tea, chesses, fish, synthetic beverages, meat, grains, fruits, and vegetables (Parra, Arrieta, Fernández-Escudero, García, et al. 2006; Parra, Arrieta, Fernández-Escudero, Íñiguez, et al. 2006; Bhattacharyya et al. 2008; Palit et al. 2010). It can also be used for environmental monitoring (toxic heavy metal analysis, water quality analysis, and plant samples). The e-tongue is a functional electronic tongue with a light source, sensor array, and detector. The light source illuminates chemically modified polymer beads placed on a sensor chip, which is a small silicon wafer. These beads change color when they enter the presence of these beads.

1.3 SENSORS FOR PROCESS IMPROVEMENT

Sensors are now widely available and reasonably priced, and more devices are on the horizon, so there is no excuse for not understanding what is going on in your process. Sensors are currently widely used (Vielberth, Englbrecht, and Pernul 2020). Proximity sensors, accelerometers, gyroscopes, magnetometers, biometric devices, infrared and temperature sensors, humidity sensors, and even oximeters are all built into your smartphone (though you may not want to depend on the accuracy of a smartphone-based oximeter). In addition to the four basic sensors that provide a window into industrial processes (flow, level, pressure, and temperature), a variety of different sensing devices may be found in daily products ranging from toys to dishwashers to automobiles. It has been opined that it may be difficult to compete for long for any food or beverage company that refuses to employ sensors to monitor their processes. The probability of controlling what is not measure zero, and without the deployment of sensors some measurements are impossible, especially in the food processing industries. Process monitoring and regulation can be achieved with the help of inventive engineering and the appropriate sensor devices. System integrators can be helpful in implementing sensor driven food process monitoring system (Wayne-Labs 2021).

1.4 THE FUTURE ROLE OF SOFTWARE IN SENSORS

Without a doubt, software is vital in today's sensing systems and will be employed in a variety of applications. Wesstrom says,

> I believe we will see a mix; some instruments will continue to require complex software to function, whilst for others, simply getting the raw signal will be sufficient. … We started with pH, conductivity, and a few other liquid analysis products, but the CM82 transmitter can be used for other measurement parameters to reduce instrument cost.
>
> Wesstrom 2014

Wesstrom claims that the food sector is fast embracing cloud-based services for asset health monitoring and traceable performance verification due to a growing need for performance and safety records. Endress+Hauser's Heartbeat Technology, which blends diagnostic, verification, and monitoring elements for process improvement, exemplifies this. According to Harvey, software is already being used in advanced diagnostic features for predictive maintenance, and maintenance engineers can use software to perform system inspections. Brooks Instrument's "smart" instruments, for example, come with onboard sophisticated diagnostics to help with predictive maintenance and troubleshooting. Although it may appear that this technology is unique to the food-processing industry, it is more prevalent than one might think. Changes to hairstyle and color, facial hair, cosmetics, sunglasses on or off, and other appearances were never preprogrammed; instead, sensors and AI technology on your phone learned and adapted to them. I am the only one who has access to my phone, regardless of how I seem. It is probably possible to adapt this to food manufacture. Perhaps the only thing that is missing right now is some inventiveness in execution. Sensors are less expensive than they were previously due to their widespread use in just about any application you can think of consumer, medical, or industrial and they have gotten smaller in general, with wireless capabilities allowing for certain previously unattainable use cases. Today's sensors are becoming more easier for process plants to install and run, thanks to AI software and built-in asset verification, and operators and management can collect critical data from their processes, and the research in these areas is still developing (Grau et al. 2021; Fu et al. 2020).

1.5 FOOD SAFETY AND QUALITY ANALYSIS USING SMARTPHONE SENSING

In 1992, the first smartphone, which contained personal digital assistant applications, was developed by International Business Machines Corporation. The early generation had limited processor and sensor performance with a short, unplugged power operation time. To date, smartphone-based sensing is still under development. Recently, several new companies have developed smartphone software for POC applications. Holomic LLC, for example, has created smartphone-based health and environmental management tools such as Allergen Tester and Mercury Analyzer. Several companies have developed smartphone software that can read and analyze data from a lateral flow test strip. Mobile Assay, for example, has created a mobile reader based on the Instantaneous Analysist program. Cooper and co-workers developed mobile image radiometry (MIR) algorithms, which were used to run the app. This app was able to convert the visually detectable colorimetric signal from the rapid diagnostic test to a digital signal and interpret the signal by comparing the color of the pixels, and the data can be stored in the secure cloud of the mobile assay. It was compatible with Windows, Android, and iOS software. Similarly, another company named Navarum has also created smartphone apps that could read the QR code integrated into lateral flow cassettes. The apps could interpret the signal by comparing the color of the pixels. Some other companies are developing smartphone camera lenses that enhance zoom and image quality, similar to benchtop microscopes. BLIP Lenses, for example, provide low-cost lenses that are compatible with any mobile phone. The lenses were designed to be attached to any phone camera. Although some commercial smartphone-based sensors have been widely used for medical diagnosis, they are applicable for the detection of various target analytes and thus may offer tremendous potential for food safety monitoring soon (Chinnapaiyan et al. 2022; Yue et al. 2022).

1.6 FUTURE PERSPECTIVE

The growing number of studies over the past few years indicates that mobile sensing is an area that is evolving extremely quickly. This section examines the most recent developments in the creation of smartphone-based sensors for food safety and quality monitoring. Smartphones can be equipped with paper-based sensors, chip-based sensors, tube or microwell assays, and colorimetric, fluorescent, luminescent, and electrochemical techniques for the detection of chemicals, toxins, and food pathogens (Franca and Oliveira 2021; Ru Choi 2017; Chinnapaiyan et al. 2022). Smartphone-based sensors for environmental, food safety, and health applications are now available for purchase. The smartphone app enables untrained users to wirelessly upload data in order to detect and quantify target analytes. View online as a stand-alone tool for the identification of certain analytes outside of the lab, particularly in underdeveloped countries.

The use of sensors on smartphones still faces various difficulties, despite the fact that this subject is still in its infancy. For instance, the present smartphone-based platform needs extra equipment, including pumps and pipette tips, to introduce the sample. A sample collection device should be integrated in subsequent work to create an all-in-one device. Additionally, the majority of sensors call for steps in off-chip sample preparation, which heavily rely on specialized labor. Therefore, creating a sample-to-answer interface would be great to do away with laborious operational procedures. To enable automated on-chip test and fluidic control, self-powered integrated sensors should be created. Furthermore, the capacity to multiplex the detection would greatly increase assay productivity. Chemical storage on the chip is essential to obviate the need for a laboratory storage unit (such as a refrigerator). Even in testing situations with extreme temperature and humidity, the ability to monitor environmental parameters on the chip (for instance, temperature and relative humidity) would lead to good assay performance.

The wireless network typically suffers from poor connectivity in environments with few resources. The platform must therefore be able to enable asynchronous data transmission in order to guarantee the reliability of smartphone-based sensor applications. The performance of sensors in rural areas with sparse power supplies would also be improved by the use of alternate power sources, such as solar power and battery technology. A normalization technique that enables optimal image collection and processing even under difficult lighting conditions needs to be developed for smartphone modules, which calls for a greater knowledge of lighting bias.

Additionally, comparisons of the consistency between various smartphone platforms are necessary to pinpoint the variations between them in terms of light sources and/or cameras in order to examine assays and enhance their capacity for precise signal measurement.

In conclusion, smartphone-based sensing is a promising area with lots of room for both scientific study and business development. Smartphone-based sensors are expected to play a significant role in the monitoring and efficient management of food contamination outbreaks in the near future.

1.7 CONCLUSION

Traditional techniques for quality and safety analysis are time-consuming and labor-intensive and involve the use of experienced personnel. Because improper handling and storage can lead to food poisoning, this approach cannot be trusted. In the food-processing industry, this can be avoided by adopting the sensor-automation technique. The capabilities, restrictions, and applicability of sensors for the intended application must be carefully considered when selecting sensors. Real-time analysis, high sensitivity, repeatability, selectivity, and mobility, cheap cost of ownership, and gradual replacement and/or parallel utilization of costly and tedious analytical laboratory equipment are just a few of the advantages that food sensors bring. Portable detection devices that are low-power, humidity resistant and cost-effective, as well as a network of sensor arrays with quick screening, monitoring, and reporting, are conceivable results. Sensor technology assists in the detection and identification of pollutants during food production activities, resulting in better food quality, safety, productivity, and profitability for the food processing sector. This chapter presented various sensor device and technologies relevant to food industries with the relevant application in food monitoring, food safety, food control, and preservation of food chain.

REFERENCES

Adetunji, C.O., Nwankwo, W., Ukhurebor, K.E., Olayinka, A.S., and Makinde, A.S., 2021. Application of biosensor for the identification of various pathogens and pests mitigating against the agricultural production: Recent advances. *Biosensors in Agriculture: Recent Trends and Future Perspectives*, 169–189. Cham: Springer International Publishing. https://doi.org/10.1007/978-3-030-66165-6_9

Arman Kuzubasoglu, B. and Kursun Bahadir, S., 2020. Flexible temperature sensors: A review. *Sensors and Actuators, A: Physical*, 315, 112282.

Balbinot, S., Srivastav, A.M., Vidic, J., Abdulhalim, I., and Manzano, M., 2021. Plasmonic biosensors for food control. *Trends in Food Science and Technology*, 111, 128–140.

Bhattacharyya, N., Bandyopadhyay, R., Bhuyan, M., Tudu, B., Ghosh, D., and Jana, A., 2008. Electronic nose for black tea classification and correlation of measurements with "Tea taster" marks. *IEEE Transactions on Instrumentation and Measurement*, 57 (7), 1313–1321.

Chinnapaiyan, S., Rajaji, U., Chen, S.M., Liu, T.Y., de Oliveira Filho, J.I., and Chang, Y.S., 2022. Fabrication of thulium metal–organic frameworks based smartphone sensor towards arsenical feed additive drug detection: Applicable in food safety analysis. *Electrochimica Acta*, 401, 139487.

Cui, X., Liu, G., and Lin, Y., 2005. Amperometric biosensors based on carbon paste electrodes modified with nanostructured mixed-valence manganese oxides and glucose oxidase. *Nanomedicine: Nanotechnology, Biology, and Medicine*, 1 (2), 130–135.

Curulli, A., 2021. Electrochemical biosensors in food safety: Challenges and perspectives. *Molecules*, 26 (10), 2940.

Emanuelli, J., Pagnussat, V., Krieser, K., Willig, J., Buffon, A., Kanis, L.A., Bilatto, S., Correa, D.S., Maito, T.F., Guterres, S.S., Pohlmann, A.R., and Külkamp-Guerreiro, I.C., 2021. Polycaprolactone and polycaprolactone triol blends to obtain a stable liquid nanotechnological formulation: Synthesis, characterization and in vitro – in vivo taste masking evaluation. *Drug Development and Industrial Pharmacy*, 47 (10), 1556–1567.

Fracchiolla, N.S., Artuso, S., and Cortelezzi, A., 2013. Biosensors in clinical practice: Focus on oncohematology. *Sensors (Switzerland)*, 13 (5), 6423–6447.

Franca, A.S. and Oliveira., L.S., 2021. Applications of smartphones in food analysis. In *Smartphone-Based Detection Devices: Emerging Trends in Analytical Techniques*, pp. 249–268.

Fu, B., Damer, N., Kirchbuchner, F., and Kuijper, A., 2020. Sensing technology for human activity recognition: A comprehensive survey. *IEEE Access*, 8, 83791–83820.

Ghindilis, A.L., Atanasov, P., Wilkins, M., and Wilkins, E., 1998. Immunosensors: Electrochemical sensing and other engineering approaches. *Biosensors and Bioelectronics*, 13 (1), 113–131.

Goud, K.Y., Moonla, C., Mishra, R.K., Yu, C., Narayan, R., Litvan, I., and Wang, J., 2019. Wearable electrochemical microneedle sensor for continuous monitoring of Levodopa: Toward Parkinson management. *ACS Sensors*, 4 (8), 2196–2204.

Grau, A., Indri, M., Lo Bello, L., and Sauter, T., 2021. Robots in industry: The past, present, and future of a growing collaboration with humans. *IEEE Industrial Electronics Magazine*, 15 (1), 50–61.

Hayat, A., Catanante, G., and Marty, J.L., 2014. Current trends in nanomaterial-based amperometric biosensors. *Sensors (Switzerland)*, 14 (12), 23439–23461.

Hua, Z., Yu, T., Liu, D., and Xianyu, Y., 2021. Recent advances in gold nanoparticles-based biosensors for food safety detection. *Biosensors and Bioelectronics*, 179 (1), 113076.

Hussein, Z.K., Hadi, H.J., Abdul-Mutaleb, M.R., and Mezaal, Y.S., 2020. Low cost smart weather station using Arduino and ZigBee. *Telkomnika (Telecommunication Computing Electronics and Control)*, 18 (1), 282–288.

Karimi-Maleh, H., Karimi, F., Alizadeh, M., and Sanati, A.L., 2020. Electrochemical sensors, a bright future in the fabrication of portable kits in analytical systems. *Chemical Record*, 20 (7), 682–692.

Kauffmann, J.M. and Guilbault, G.G., 1991. Potentiometric enzyme electrodes. *Bioprocess Technology*, 15, 63–82.

Kayumba, P.C., Huyghebaert, N., Cordella, C., Ntawukuliryayo, J.D., Vervaet, C., and Remon, J.P., 2007. Quinine sulphate pellets for flexible pediatric drug dosing: Formulation development and evaluation of taste-masking efficiency using the electronic tongue. *European Journal of Pharmaceutics and Biopharmaceutics*, 66 (3), 460–465.

Kim, J., Cho, J.H., Lee, H.M., and Hong, S.M., 2021. Capacitive humidity sensor based on carbon black/polyimide composites. *Sensors*, 21 (6), 1974.

Komazaki, Y. and Uemura, S., 2019. Stretchable, printable, and tunable PDMS-CaCl2 microcomposite for capacitive humidity sensors on textiles. *Sensors and Actuators, B: Chemical*, 297, 126711.

Kovalenko, O. and Chuprina, R., 2022. Machine learning and AI in food industry solutions and potential [online].

Kumar, H. and Rani, N., 2016. Enzyme-based electrochemical biosensors for food safety: A review. *Nanobiosensors in Disease Diagnosis*, 5, 29–39.

Kurbanoglu, S., Erkmen, C., and Uslu, B., 2020. Frontiers in electrochemical enzyme based biosensors for food and drug analysis. *TrAC – Trends in Analytical Chemistry*, 124, 115809.

Law Al, A.T. and Adeloju, S.B., 2013. Progress and recent advances in phosphate sensors: A review. *Talanta*, 114, 191–203.

Li, F., Yu, Z., Han, X., and Lai, R.Y., 2019. Electrochemical aptamer-based sensors for food and water analysis: A review. *Analytica Chimica Acta*, 1051, 1–23.

Liang, R., Luo, A., Zhang, Z., Li, Z., Han, C., and Wu, W., 2020. Research progress of graphene-based flexible humidity sensor. *Sensors (Switzerland)*, 20 (19), 5601.

Lin, C.H., Lin, J.H., Chen, C.F., Ito, Y., and Luo, S.C., 2021. Conducting polymer-based sensors for food and drug analysis. *Journal of Food and Drug Analysis*, 29 (4), 544–558.

Lv, C., Hu, C., Luo, J., Liu, S., Qiao, Y., Zhang, Z., Song, J., Shi, Y., Cai, J., and Watanabe, A., 2019. Recent advances in graphene-based humidity sensors. *Nanomaterials*, 9 (3), 422.

Lv, M., Liu, Y., Geng, J., Kou, X., Xin, Z., and Yang, D., 2018. Engineering nanomaterials-based biosensors for food safety detection. *Biosensors and Bioelectronics*, 106, 122–128.

Mehrotra, P., 2016. Biosensors and their applications – A review. *Journal of Oral Biology and Craniofacial Research*, 6 (2), 153–159.

Misra, N.N., Dixit, Y., Al-Mallahi, A., Bhullar, M.S., Upadhyay, R., and Martynenko, A., 2020. IoT, big data and artificial intelligence in agriculture and food industry. *IEEE Internet of Things Journal*, 4662 (c), 1–1.

Mohan, H.K.S.V., Toh, C.H., and Malcolm, A.A., 2021a. Non-destructive evaluation of Flexible Sachet Seal Integrity using a capacitive proximity sensor. In: *IEACon 2021–2021 IEEE Industrial Electronics and Applications Conference*, pp. 169–174.

Mohan, H.K.S.V., Toh, C.H., and Malcolm, A.A., 2021b. Non-destructive evaluation of Container Capping Integrity using capacitive proximity sensor. In: *2021 5th International*

Conference on Electronics, Materials Engineering and Nano-Technology, IEMENTech 2021, pp. 1–5.

Moustafa, H., Morsy, M., Ateia, M.A., and Abdel-Haleem, F.M., 2021. Ultrafast response humidity sensors based on polyvinyl chloride/graphene oxide nanocomposites for intelligent food packaging. *Sensors and Actuators, A: Physical*, 331, 112918.

Mulchandani, P., Mulchandani, A., Kaneva, I., and Chen, W., 1999. Biosensor for direct determination of organophosphate nerve agents. 1. Potentiometric enzyme electrode. *Biosensors and Bioelectronics*, 14 (1), 77–85.

Naresh, V. and Lee, N., 2021. A review on biosensors and recent development of nanostructured materials-enabled biosensors. *Sensors (Switzerland)*, 21 (4), 1109.

Nguyen, T.D., Kim, T., Noh, J., Phung, H., Kang, G., and Choi, H.R., 2021. Skin-type proximity sensor by using the change of electromagnetic field. *IEEE Transactions on Industrial Electronics*, 68 (3), 2379–2388.

Nooriman, W.M., Abdullah, A.H., Rahim, N.A., and Kamarudin, K., 2018. Development of wireless sensor network for Harumanis Mango orchard's temperature, humidity and soil moisture monitoring. In: *ISCAIE 2018–2018 IEEE Symposium on Computer Applications and Industrial Electronics*, pp. 263–268.

Nwankwo, W., Olayinka, A.S., and Ukhurebor, K.E., 2019. The urban traffic congestion problem in Benin city and the search for an ICT-improved solution. *International Journal of Scientific and Technology Research*, 8 (12), 65–72.

Olayinka, A.S., Adetunji, C.O., Nwankwo, W., Olugbemi, O.T., and Olayinka, T.C., 2022. A study on the application of Bayesian learning and decision trees IoT-enabled system in postharvest storage. In: S. Pal, D. De, and R. Buyya, eds. *Artificial Intelligence-based Internet of Things Systems*. Cham: Springer International Publishing, pp. 467–491.

Palit, M., Tudu, B., Dutta, P.K., Dutta, A., Jana, A., Roy, J.K., Bhattacharyya, N., Bandyopadhyay, R., and Chatterjee, A., 2010. Classification of black tea taste and correlation with tea taster's mark using voltammetric electronic tongue. *IEEE Transactions on Instrumentation and Measurement*, 59 (8), 2230–2239.

Parra, V., Arrieta, Á.A., Fernández-Escudero, J.A., García, H., Apetrei, C., Rodríguez-Méndez, M.L., and Saja, J.A. d., 2006. E-tongue based on a hybrid array of voltammetric sensors based on phthalocyanines, perylene derivatives and conducting polymers: Discrimination capability towards red wines elaborated with different varieties of grapes. *Sensors and Actuators, B: Chemical*, 115 (1), 54–61.

Parra, V., Arrieta, Á.A., Fernández-Escudero, J.A., Íñiguez, M., Saja, J.A. De, and Rodríguez-Méndez, M.L., 2006. Monitoring of the ageing of red wines in oak barrels by means of an hybrid electronic tongue. *Analytica Chimica Acta*, 563 (1–2), 229–237.

Popa, A., Hnatiuc, M., Paun, M., Geman, O., Hemanth, D.J., Dorcea, D., Son, L.H., and Ghita, S., 2019. An intelligent IoT-based food quality monitoring approach using low-cost sensors. *Symmetry*, 11 (3), 374.

Pradhan, S., Jityen, A., Juagwon, T., Sinsarp, A., and Osotchan, T., 2020. Development of electrochemical electrodes using carbon nanotube and metal phthalocyanine to classify pharmaceutical drugs. *Materials Today: Proceedings*, 23, 732–737.

Ramanathan, K. and Danielsson, B., 2001. Principles and applications of thermal biosensors. *Biosensors and Bioelectronics*, 16 (6), 417–423.

Rohit, C., Moos, M., Meldrum, R., and Young, I., 2019. Comparing infrared and probe thermometers to measure the hot holding temperature of food in a retail setting. *Food Protection Trends*, 39 (1), 74–83.

Ru Choi, J. 2017. Chapter 11: Smartphone-based sensing in food safety and quality analysis. In Food Chemistry, Function and Analysis, Vol. 2017, January. https://doi.org/10.1039/9781788010528-00332.

Sadrieh, N., Brower, J., Yu, L., Doub, W., Straughn, A., MacHado, S., Pelsor, F., Saint Martin, E., Moore, T., Reepmeyer, J., Toler, D., Nguyenpho, A., Roberts, R., Schuirmann, D.J., Nasr, M., and Buhse, L., 2005. Stability, dose uniformity, and palatability of three counterterrorism drugs – Human subject and electronic tongue studies. *Pharmaceutical Research*, 22 (10), 1747–1756.

Shun, W.G., Muda, W.M.W., Hassan, W.H.W., and Annuar, A.Z., 2020. Wireless sensor network for temperature and humidity monitoring systems based on NodeMCU ESP8266. In: *Communications in Computer and Information Science*, pp. 1132, 262–273.

Sinambela, R., Mandang, T., Subrata, I.D.M., and Hermawan, W., 2020. Application of an inductive sensor system for identifying ripeness and forecasting harvest time of oil palm. *Scientia Horticulturae*, 265, 109231.

Spink, C. and Wadsö, I., 1976. Calorimetry as an analytical tool in biochemistry and biology. *Methods of Biochemical Analysis*, 23 (0), 1–159.

Steinberg, C., Philippon, F., Sanchez, M., Fortier-Poisson, P., O'Hara, G., Molin, F., Sarrazin, J.F., Nault, I., Blier, L., Roy, K., Plourde, B., and Champagne, J., 2019. A novel wearable device for continuous ambulatory ECG recording: Proof of concept and assessment of signal quality. *Biosensors*, 9 (1), 17.

Su, Y., Ma, C., Chen, J., Wu, H., Luo, W., Peng, Y., Luo, Z., Li, L., Tan, Y., Omisore, O.M., Zhu, Z., Wang, L., and Li, H., 2020. Printable, highly sensitive flexible temperature sensors for human body temperature monitoring: A review. *Nanoscale Research Letters*, 15 (1), 200.

Thakur, M.S. and Ragavan, K. V., 2013. Biosensors in food processing. *Journal of Food Science and Technology*, 50 (4), 625–641.

Vielberth, M., Englbrecht, L., and Pernul, G., 2020. Improving data quality for human-as-a-security-sensor. A process driven quality improvement approach for user-provided incident information. *Information and Computer Security*, 29 (2), 332–349.

Villalonga, A., Sánchez, A., Mayol, B., Reviejo, J., and Villalonga, R., 2022. Electrochemical biosensors for food bioprocess monitoring. *Current Opinion in Food Science*, 43, 18–26.

Wang, J., 1999. Amperometric biosensors for clinical and therapeutic drug monitoring: A review. *Journal of Pharmaceutical and Biomedical Analysis*, 19 (1–2), 47–53.

Wayne-Labs, 2021. https://digitaledition.foodengineeringmag.com/february2021/tech-update-sensors/ (Accessed 27 May 2022).

Wesstrom, O., 2014. Supplementing lab analysis with inline measurements. *Control Engineering*, 61 (1).

Wu, Q., Zhang, Y., Yang, Q., Yuan, N., and Zhang, W., 2019. Review of electrochemical DNA biosensors for detecting food borne pathogens. *Sensors (Switzerland)*, 19 (22), 4916.

Ye, Y., Zhang, C., He, C., Wang, X., Huang, J., and Deng, J., 2020. A review on applications of capacitive displacement sensing for capacitive proximity sensor. *IEEE Access*, 8, 45325–45342.

Yue, X., Li, Y., Xu, S., Li, J., Li, M., Jiang, L., Jie, M., and Bai, Y., 2022. A portable smartphone-assisted ratiometric fluorescence sensor for intelligent and visual detection of malachite green. *Food Chemistry*, 371, 131164.

Yunus, G. and Kuddus, M., 2020. Electrochemical biosensor for food borne pathogens: An overview. *Carpathian Journal of Food Science and Technology*, 12 (2), 5–6.

Zadeh, S. M. H., Heidarshenas, M., Ghalambaz, M., Noghrehabadi, A., and Pour, M. S., 2020. Numerical modeling and investigation of amperometric biosensors with perforated membranes. *Sensors (Switzerland)*, 20 (10), 118–124. https://doi.org/10.3390/s20102910

Zheng, J.Y. and Keeney, M.P., 2006. Taste masking analysis in pharmaceutical formulation development using an electronic tongue. *International Journal of Pharmaceutics*, 310 (1–2).

2 Sensing and Artificial Intelligence for Food Packaging

Olorunsola Adeyomoye, Charles Oluwaseun Adetunji, Olugbemi T. Olaniyan, Akinola Samson Olayinka, and Olalekan Akinbo

CONTENTS

2.1 INTRODUCTION

Sensing and artificial intelligence are two key domains in achieving the goal of creating a high degree of accuracy in food packaging (Adetunji et al., 2020; Adetunji et al., 2020a,b,c; Oyedara et al., 2022; Adetunji et al., 2022a,cb;d,e,f,g,h,i; Olaniyan et al., 2022a,b). The integration of these two domains allows for better data collection and design of biosensors for surveillance of environmental conditions that could result in food spoilage and decay (Jin, 2020). Important components of artificial intelligence-biosensor technology are data processing, signal acquisition and innovation, biorecognition components, and intelligent decision making, all of which are important in redefining food packaging and processing for prolongation of shelf life and food availability for most people (Tittl, 2019).

New molecular biosensing techniques based on the combination of dielectric metasurface and image detection utilize functional spectrophotometer. Food processing and packaging technologies – including drying, high temperature processing, irradiation, 3D food printing and microwave sterilization – have significantly improved over time. The current food system utilizes biopolymers for analysis as well

DOI: 10.1201/9781003207955-2

as the packaging of food substances (Jiang, 2020). Biosensors are readily available to monitor protein-based food products to prevent their spoilage (Rogers, 2016).

These biosensors operate by detecting changes in color in pH indicators via halochromic behavior. In contrast, when food is treated chemically, some of these sensors may be altered, thereby resulting in reversible changes in the true history and quality of spoiled food products. Biosensors could be made use of as non-toxic film sensors showing halochromic behavior in bioamine supplied by contaminated seafood and meat products (Liu, 2020). These biosensors do not undergo changes in color following bioamine release, therefore sensors can be used as reliable anti-protein sensor for protein food products.

2.2 ADVANCES IN FOODBORNE PATHOGENS

Diseases resulting from exposure to contaminated food products have been of major public health concern with 48 million morbidity and 3,000 deaths yearly (Barnes, 2022). Fruits and vegetables have been the major sources of many of these diseases. Epidemiologic studies shows 20 percent outbreaks related to the diseases. *Shiga toxin Escherichia coli, Campylobacter, Salmonella, Shigella, and Cryptosporidium*, have continued to be major causes of foodborne laboratory infections, according to reports of Foodborne Diseases Active Surveillance Network (Gallo, 2020).

The contamination of food along the food chain can occur at any stage, that is, on the farm or in the consumer's kitchen. Therefore, it is important to put in place control measures to reduce the burden and spread of foodborne illness (Choi, 2019). New agents that create life-threatening conditions are constantly being identified, with an increase in the number of outbreaks. These diseases continue to have a profound effect on children, adults, and their immune systems.

Salmonella is still the most common cause of severe food poisoning; in fact, *Salmonella typhimurium* and *Salmonella enteritidis* species are very dangerous. Foodborne disease occurs frequently in restaurants and homes, and the major causes of the outbreak are lack of temperature control for the prepared food, hygiene during cooking, and the food. In developed countries, *Clostridium botulinum,* VTEC and Salmonella, are more common than in developing countries.

In developing nations, *Vibrio cholerae*, *Shigella*, and parasites are the most prevailing diseases; it is not clear how many cases are related to food, water, or human-to-human transmission. Temperature control in kitchens could be associated with a clear decrease in *C. perfringens* and *S aureus* in developed countries. Rapid awareness and coordination against Salmonella have been promoted by Salm-Net and many other international organizations. However, if such networks do not exist worldwide, disease surveillance groups may likely have very limited facilities for research in the developing world.

Several developed nations have invested in the control and monitoring of foodborne diseases, but this same commitment is needed by the World Health Organization. In order to prevent foodborne disease, measures should be put in place in key areas of the food supply chain, from food production to preparation. Microorganisms and food items that result in major outbreaks of diseases should be the focus of efforts in preventing foodborne disease. Over many years, *Listeria monocytogenes*, a cell-rich bacterium, has been associated with the number of diseases found in food (Farber, 1991). It is a virus that leads to abortions, neonatal deaths, septicemia, and meningitis in pregnant women, their neonates and adults with dementia. It results in symptoms that include abortion, neonatal death, septicemia, and meningitis. The bacterium can be seen in a variety of foods, with a strong focus on meat, poultry, and marine animals.

There is progress in the technology used for the identification and quantification of microorganisms in food, such as those based on DNA probes, monoclonal antibodies, and the polymerase chain reaction. In addition, progress in preventing and controlling human infection has been made as much as the molecular knowledge and applied biology of *L. monocytogenes* continue to emerge. Staphylococcal-borne disease remains one of the major cause of foodborne illnesses in the world, caused by *Staphylococcus aureus* enterotoxins in the diet (Kadariya, 2014). It is one of the most commonly reported foodborne illnesses.

Although many Staphylococcal enterotoxins have been identified, the most common cause of this disease worldwide is *Staphylococcus aureus*, which is resistant to heat. Most outbreaks are caused by poor food handling processes in the retail business, according to an outbreak investigation (Wendlandt, 2013). However, many studies have reported that *S. aureus* is very common in a variety of food products, including raw meat for sale, which means consumers may be at risk of *S. aureus* colonization and infection. The presence of germs in food products is dangerous to consumers, leading to significant economic and product losses. Abdominal pain, nausea, and vomiting are all symptoms of *S. aureus*. Food safety and preparation, cold chain maintenance, adequate cleaning and disinfection of equipment, prevention of home and kitchen contamination, and prevention of contamination from one farm to another are examples of food-safety measures.

2.3 ENVIRONMENT AND FOOD SAFETY

Diet has undergone major changes in recent years, and now food is said to play a crucial role in the maintenance of good physical and mental health, as well as in preventing certain diseases, in addition to your nutritional and sensory features (Adetunji et al., 2021a,b,c; Ukhurebor et al., 2021a,b; Sangeetha et al., 2021; Adetunji and Ukhurebor, 2021, Oluwaseun et al., 2017; Adetunji et al., 2012; Arowora et al., 2012; Dauda et al., 2022a,b; Okeke et al., 2021; Adetunji and Anani, 2021; Nwankwo et al., 2021; Adejumo and Adetunji. 2018). Eating can contain a lot of health risks. Many processes that are important in food production can create real risks of contamination or unsafe food production for consumers. Maintaining the safety of food products, proper storage, physicochemical and microbiological stability, and cooking methods are some of the important ways of maintaining the safety of food products (Chapman, 2018).

Despite the fact that food safety testing is an important health protection strategy used by governments to prevent foodborne illness, many have continued to criticize this measure. Lack of compliance with the screening procedures, and inefficiency in preventing foodborne illness are among the factors contributing to defective food safety procedures. Assessing the validity of this criticism is an important topic for future research. In addition, food practitioners and consumers require better food-safety education programs. Food security measures such as mass production and distribution of food, global trade in food security, home-based food, the introduction of newborn diseases, and the growing number of at-risk consumers should all be addressed (Nyachuba, 2010). The modern risk-based food security system adopts a fast-paced approach to food security from one farm to another, based on busy data collection and analysis to better identify potential risks and risk factors, design and evaluate treatment, and prioritize prevention efforts. A strategy like this focuses on limited resources in parts of the food chain that are likely to improve public health.

Outbreaks of foodborne illness are becoming increasingly common as community kitchens, restaurants, and local community-based food systems, such as agriculture, become more common (Todd, 1997). Traditional methods of detecting food contamination, such as fluid chromatography and mass spectrometry, have previously been used, but are rather expensive, time-consuming, and labor-intensive, limiting their use in landscaping programs. Machines, such as paper-based and chip-based devices, are generally faster, less expensive, and easier to use, and they have greater potential for food-safety analysis (Gould, 2013).

Raman spectroscopy is a recent technique used in assessing food safety in a safe, easy-to-use, sensitive, and fast way. Recent advances in Raman spectroscopy technologies have greatly improved the ability to detect food contaminants, resulting in a dramatic increase in its use in food safety (Petersen, 2021). Pre-harvest food security is related to the great scientific achievements that are the product of collaborative efforts and the results of remarkable research. Biological production, antibiotic resistance, genomic sequence, and functional metrics are among the pre-harvest food safety topics that draw attention to potential food safety challenges and future food safety research programs (Torrence, 2016). Foodborne pathogens, hazardous substances, and scientific and policy interactions may continue to be the site of foodborne pathogens, hazardous substances in food safety research, whether standard or before harvest. Priorities and research in the field of food safety should be accompanied by new global concerns, technologies, and approaches based on a multi-sectoral, collaborative, and systematic approach.

2.4 FOOD PACKAGING MATERIALS

Food packaging is a crucial factor in maintaining food quality supplies. It performs a crucial role in keeping the product packaged from the elements, preserving food quality, and improving the qualities of the packaged food during storage. However, the regular packaging materials are made of polymers that do not decay. As a result, many recent studies have paid attention to developing packaging materials based on biopolymer. Biopolymers can be considered as active packaging materials because they can improve the transport of a wide variety of active chemicals. To improve food quality and safety, new food packaging techniques are an important research topic.

Organic and edible packaging have become very popular. Edible packaging uses a continuous, perishable food item as a useful or adhesive material (Trajkovska Petkoska,

2021). Many recent studies have tested the usefulness of dietary supplements as an additional value in packaged foods. Nanotechnology has evolved as an alternative way to combine vitamins, antimicrobials, bioactives, nutrients, and antioxidants into edible packages to improve efficiency (Kraśniewska, 2020). It is used to disperse nutrients that include nanofibers, nanoemulsions, encapsulants, and nanoparticles, in an edible form.

Edible packaging serves as an effective packaging in this way and plays a crucial role in food packaging in conjunction with nutritious food and performing technical functions such as releasing nutrients such as antibiotics and antioxidants that improve the shelf life of food products. Functional food packaging may also assist in maintaining the profile of packaged foods. The inclusion of anti-bacterial agents in packaged foods is at the forefront of research and modern development of food packaging (Gumienna, 2021). Disposable chitosan and biocompatible is one of the newest natural polymers on the market with antimicrobial activities. Producers and scientists from many fields have developed an interest in producing chitosan alone or as a composite film based on chitosan because of its properties. Chitosan films have the potential to be packaged to keep food fresh and pathogen safe. Furthermore, chitosan is often used in antimicrobial films to protect against various pathogenic bacteria and food. Antimicrobial agents can be applied to packaging materials by directly bonding antimicrobials to polymers or by coating polymer surfaces with antiseptic coatings (Fu, 2020).

Another type is thermal antimicrobials, often accomplished with thermal filming technologies, including compression molding or film extraction. Extra coverage, which is a non-heating method, is a more promising method than molding or dispensing in producing food packaging materials that include antimicrobial resistance to heat. It provides the advantage of maintaining bulk packaging properties like mechanical and physical strength as well as reducing the amount of antimicrobials needed to achieve optimal performance over direct combinations. Polylactic acid is often regarded as one of the most promising and environmentally friendly polymers due to its physical and chemical properties, including biological decomposition, regeneration, biodiversity, and safety. The study of polylactic acid as another food packaging film with enhanced features has increased its usefulness in food packaging. Figure 2.1 shows the relationship between artificial intelligence and sensors technology that are applicable in packaging.

2.5 DETECTION OF FOOD PATHOGENS USING BIOSENSING TECHNOLOGIES

In recent years, global biosecurity challenges like the spread of new infectious agents and bioterrorism have aroused interest in the use of biosensors (Pejcic, 2006). Efforts are being made to diagnose and prevent the spread of infectious diseases. Biosensors are attractive tools that have the power to detect disease outbreaks. Despite the existence of various modern technologies, which have been reported in scientific literature, the development of neurological diagnostics is still relatively new (Pejcic, 2006). There is no doubt that glucose biosensors, gene chips, protein chips, and other

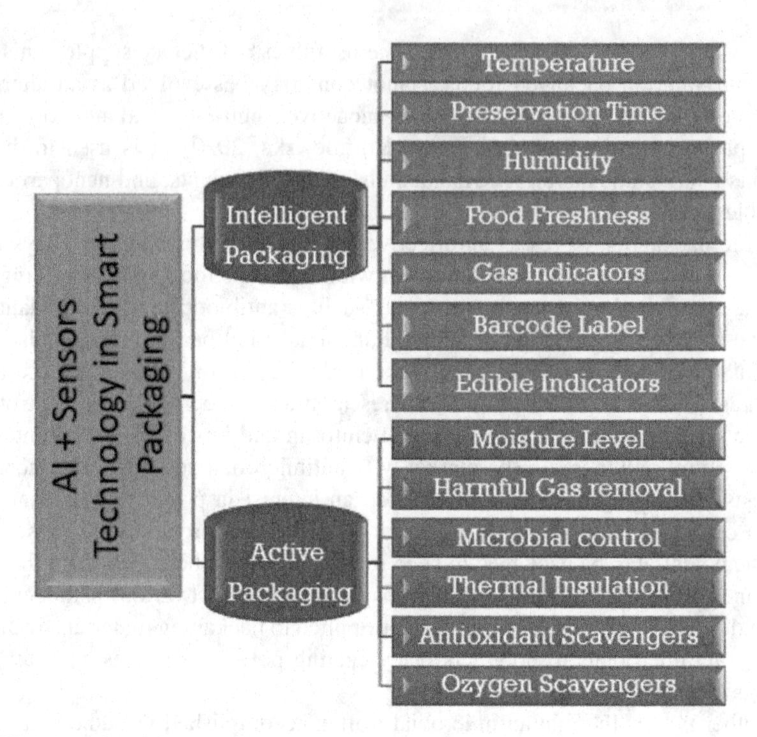

FIGURE 2.1 AI and Sensors Technology in Packaging.

similar devices continue to play an important role in monitoring the abundance of biomolecules. Infections caused by bacteria, viruses, and parasite are a global health concern, and are also present in contaminated drinking water (Connelly, 2012).

Biosensors based on a range of biorecognition molecules and transfer mechanisms have been reported, capable of providing the most sensitive and direct detection of interest in the short term (Cheng, 2013). Immunosensors provide signal-based stimuli through the use of antibodies or antigens as a unique sensor (Jiang, 2008; Palchetti, 2009). Immunosensing technology has a lot of potential for fact detection of pesticide residues in food and the environment. These various transfer technology systems include piezoelectric, electrochemical, optical, and nanomechanic systems, all of which have been documented in the design and production of pesticide diagnostic tests.

It utilizes a variety of configuration processes to provide a visual biorecognition connector. In addition, in the development and use of immunosensors, mechanisms such as signal amplification, miniaturization, regeneration, and antibodies have been tested. Despite some of the limitations in the use of immunosensors, they are increasingly critical in environmental and analysis of dietary food products. Another product developed from nanotechnology are nanosensors used in food processing and packaging. These novel biosensors are often considered better ways of determining food contaminants than the traditional methods. Many food contaminants can be accurately detected using nanotechnology and nanominerals (Li, 2014).

2.6 BIO-BASED SENSOR FOR FOOD PACKAGING

Smart food packaging is a new technology that can monitor food quality and safety throughout the shelf life. This method uses packed sensors to monitor changes in diet following chemical and microbial degradation (Rodrigues, 2021). These indicators often provide information, including the packaging level of the packaged product, and the color change that can be easily obtained by both food distributors and consumers. However, most indicators that are currently in use are non-renewable and non-perishable synthetic materials. Because of the pressing need to increase the stability of food packaging, the selection of biosensors is crucial to this need.

Some of the bioactive ingredients found in natural extracts can be used in bio-based sensors to be used in packaging clever or intelligent foods when combined with biopolymers. Bioactive extracts, such as anthocyanins, are obtained from different sources, which include food industry products, and are potent for use as a biosensor material. The time-bound component of a smart food packaging device, is based on a glucose biosensor (Mijanur Rahman, 2018). The three-electrode potentiostat of this device produced an electrical signal and color enhancement in a biosensor system, which indicates glucose oxidase, glucose, and pH index (Neethirajan, 2018). These indices suggest that the electrical signal and color enhancement are consistent with predicting the quality of food. Sequences of unique coronavirus events were detected during packing and transporting food in the cold chain, making it necessary to use molecular diagnostics for food processing, packaging, transportation, and other communications. A fast-paced approach that can respond to the complexity of food security and diversity is urgently needed.

The new method is a biosensor molecular analysis technology based on CRISPR-Cas12a. It can detect germs, bacteria found in food, small molecules, and other things in a systematic way. From processing to packaging to transportation, the risk of the novel coronavirus attacks every step of the food business. The CRISPR-Cas12a-based biosensor analysis method offers many possibilities for detecting the number of infectious viruses (Mao, 2022). CRISPR-Cas12a can efficiently identify the presence of certain target nucleic acids and subsequent small changes, which are very useful in identifying nucleic acids and detecting viruses. In addition, the CRISPR-Cas12a system can be modified and integrated daily to detect various viruses, making it useful in diagnosing nucleic acid in the food safety industry. Despite the fact that CRISPR-Cas12a technology provides significant acquisition benefits, there are a number of technological barriers that need to be overcome in the transformation of emerging technologies into practical applications. Recent CRISPR-Cas12a-based applications based on CRISPR-Cas12a promote the development of a wide range of diagnostic solutions and have excellent promise for medical diagnosis, environmental monitoring, and dietary analysis (Adley, 2014).

2.7 ARTIFICIAL NEURAL NETWORK-BASED PREDICTIVE MODELS FOR BACTERIA GROWTH AND FOOD PACKAGING

Artificial neural network model calculations were used to quantify the specific growth rate of *Lactobacillus lambda* phase from bacterial growth data in genetically modified cooked beef products (Lou and Nakai, 2001). The predictive accuracy of this neural network was high compared other models used. The function of water, temperature, and the stable interaction of CO_2 melted with lambda were complex, using linear or second order relationships. An artificial neural network was utilized to calculate the output sensitivity of the model at each input parameter. Antibiofilm amylase activity has been shown against gram positive biofilms and gram negative bacterial strains. *P. aeruginosa* and *S. aureus* biofilms were reported to be blocked by slightly purified amylase from *Bacillus subtilis* (Lahiri, 2021). The effectiveness of the amylase antibiofilms was shown to be greater than that of amylase obtained commercially in spectrophotometric and microscopic tests.

The use of response surface methodology and synthetic sensory networks has been shown to be helpful in predicting the best conditions to eliminate biofilms. The docking interaction between amylase with extracellular polymeric substances of *S. aureus* and *P. aeruginosa* suggests that there may be an automatic power-driven mechanism. Arcobacter is a relatively new foodborne illness that has grown significantly in recent years. The emergence of Arcobacter species in the agro-ecosystem may likely to be underestimated due to limitations on existing findings and methods of classifying characters. Because Arcobacter is considered a food-and-water virus and a potential zoonotic agent, it is important to identify it quickly and accurately.

Raman spectroscopy is a quickly and easy technique used to distinguish viruses using Raman patterns that disperse cells. Confocal micro Raman spectroscopy combined with neural networks was used to detect Arcobacter viruses (Wang, 2020). A total of 82 references and separate areas for 18 Arcobacter species were included, with clinical, natural, and agri-food sources. Raman spectral reproducibility was not affected by bacterial planting length or growth rate. Using key component analysis, the Arcobacter species was successfully isolated from the closely related generation of *Campylobacter spp* and *Helicobacter spp*. Using Raman spectroscopy in combination with convolutional neural network, 97.2 percent accuracy was achieved with Arcobacter detection at the level of all the 18 Arcobacter species.

In addition, a fully integrated Raman spectroscopy-based sensory network was designed to measure the exact amount of Arcobacter type given to a bacterial compound containing biomass from 5–100 percent. Compared with native chemometrics, the use of both fully integrated neural networks has improved the accuracy of Raman spectroscopy in determining the type of bacteria. Within one hour of planting, this novel approach allows for the rapid identification and determination of Arcobacter species. The ability to quickly diagnose diseases is crucial to developing an early-warning system and in conducting epidemiological research. When Raman spectroscopy is combined with machine learning, Arcobacter is quickly and accurately detected at the level of animal species, making it an effective strategy for diagnosing bacterial infections in the food chain.

Artificial neural networks as a predictable model for testing the function of bacteria constitute an attractive technology, especially in the analysis of the Internet in the food industry, because of its accuracy and speed. Data were obtained from a research study of the interaction effects of different pH levels, water activity, and concentration of allyl isothiocyanate at different times during sausage production in lowering *E. coli* O157: H7 population (Palanichamy, 2008). The regression neural network, and artificial neural network type, and the polynomial regression line mathematical method were used to create speculative models.

Different mathematical uses were used to compare the predicted error of both models. Compared to mathematical model predictions, neural network training predictions and test data sets had fewer significant errors. The regression network models were somewhat better than the training mathematical model and test sets. When an approved production set was developed by translation, regression neural network accurately predicted the level of allyl isothiocyanate required. Artificial neural network models may also be useful for industry use in the production of boiled sausages in order to reduce the risk of E. coli O157: H7 and allowing the production of safe products that do not change because they are easy to produce, fast, and accurate. Because of the benefits of self-study, flexibility, robustness, and high durability in the map of vague or flexible matriculation structures, artificial intelligence technology has been utilized widely to solve problems of indirect performance measurement, pattern detection, data interpretation, preparation, diagnostics, control, filtering data, integration, and noise reduction in various food packaging technologies (Sun, 2019).

Quality control is an important process ensuring that only items that meet predetermined quality standards reach the end user or the next step of the production process. Many items in the food business are still being explored by workers in the packaging phase (Huang, 2007). Artificial intelligence and computer vision tools can be used to automate the process and improve efficiency and effectiveness. This default is difficult because it requires specialized solutions for the application location. A deep learning-based system can be developed to automatically control the quality of matrix-based food packaging while maintaining productivity and ensuring hundred percent asset testing (Patnode, 2022).

2.8 CONCLUSION AND FUTURE PERSPECTIVES

Synthetic sensory networks are computer-based mathematical tools inspired by the basic cell of the nervous system, a neuron. They create a simplified image of the human brain that combines parts of similar nerve endings such as living neurons. They are able to store large amounts of test information that can be used for normal operation with the help of a suitable predictive model. Artificial neural networks have proved to be useful in a variety of biological, medical, economic and climate disciplines, and in agricultural food science and technology. The food supply chain is a fast-growing integrated sector and encompasses all aspects from one farm to another, including production, packaging, distribution, storage, and continuous processing or cooking for consumption. Smart packaging contributes to the quality, safety, and food stability. In addition, packaging systems have evolved to be intelligent with the integration of emerging electronic and wireless communications and cloud data solutions. Smart food packaging is usually designed to assess the food itself and the environment, as well as the interaction to provide customers with food quality and safety information through various signals.

REFERENCES

Adejumo, I.O., & Adetunji, C.O. (2018). Production and Evaluation of Biodegraded Feather Meal using Immobilised and Crude Enzyme from *Bacillus subtilis* on Broiler Chickens. *Brazilian Journal of Biological Sciences*, *5*(10), 405–416.

Adetunji, C.O., & Anani, O.A. (2021). Bioaugmentation: A Powerful Biotechnological Techniques for Sustainable Ecorestoration of Soil and Groundwater Contaminants. In: Panpatte, D.G., Jhala, Y.K. (eds), *Microbial Rejuvenation of Polluted Environment. Microorganisms for Sustainability*, vol. 25. Springer, Singapore. https://doi.org/10.1007/978-981-15-7447-4_15

Adetunji, C.O., Anani, O.A., Olaniyan, O.T., Inobeme, A., Olisaka, F.N., Uwadiae, E.O., Obayagbona, O.N. (2021a). Recent Trends in Organic Farming. In: Soni, R., Suyal, D.C., Bhargava, P., Goel, R. (eds), *Microbiological Activity for Soil and Plant Health Management*. Springer, Singapore. https://doi.org/10.1007/978-981-16-2922-8_20

Adetunji, C.O., Egbuna, C., Oladosun, T.O., Akram, M., Michael, O., Olisaka, F.N., Ozolua, P., Adetunji, J.B., Enoyoze, G.E., & Olaniyan, O. (2021b). Efficacy of Phytochemicals of Medicinal Plants for the Treatment of Human Echinococcosis. Ch 8. In: *Neglected Tropical Diseases and Phytochemicals in Drug Discovery*. Wiley. DOI: 10.1002/9781119617143

Adetunji, C.O., Michael, O.S., Nwankwo, W., Ukhurebor, K.E., Anani, O.A., Oloke, J.K., Varma, A., Kadiri, O., Jain, A., & Adetunji, J.B.. (2021c). Quinoa, The Next Biotech Plant: Food Security and Environmental and Health Hot Spots. In: Varma, A. (eds), *Biology and Biotechnology of Quinoa*. Springer, Singapore. https://doi.org/10.1007/978-981-16-3832-9_19

Adetunji, C.O., Mitembo, W.P., Egbuna, C., & Narasimha Rao, G.M. (2020). Silico Modeling as a Tool to Predict and Characterize Plant Toxicity. In: Mtewa, A.G., Egbuna, C., Narasimha Rao, G.M. (eds), *Poisonous Plants and Phytochemicals in Drug Discovery*. Wiley Online Library. https://doi.org/10.1002/9781119650034.ch14.

Adetunji, C.O., Nwankwo, W., Olayinka, A.S., Olugbemi, O.T., Akram, M., Laila, W., Samuel, M.O., Oshinjo, A.M., Adetunji, J.B., Okotie, G.E., & (Diuto) Esiobu, N. (2022a). Computational Intelligence Techniques for Combating COVID-19. In: *Medical Biotechnology, Biopharmaceutics, Forensic Science and Bioinformatics*. CRC Press, p. 12. eBook ISBN 9781003178903. DOI: 10.1201/9781003178903-16

Adetunji, C.O., Olaniyan, O.T., Adeyomoye, O., Dare, A., Adeniyi, M.J., Alex, E., Rebezov, M., Garipova, L., Ali Shariati, M. (2022b). eHealth, mHealth, and Telemedicine for COVID-19 Pandemic. In: Pani, S.K., Dash, S., dos Santos, W.P., Chan Bukhari, S.A., & Flammini, F. (eds), *Assessing COVID-19 and Other Pandemics and Epidemics using Computational Modelling and Data Analysis*. Springer, Cham. https://doi.org/10.1007/978-3-030-79753-9_10

Adetunji, C.O., Olaniyan, O.T., Adeyomoye, O., Dare, A., Adeniyi, M.J., Alex, E., Rebezov, M., Petukhova, E., Ali Shariati, M. (2022c). Machine Learning Approaches for COVID-19 Pandemic. In: Pani, S.K., Dash, S., dos Santos, W.P., Chan Bukhari, S.A., Flammini, F. (eds). *Assessing COVID-19 and Other Pandemics and Epidemics using Computational Modelling and Data Analysis*. Springer, Cham. https://doi.org/10.1007/978-3-030-79753-9_8

Adetunji, C. O., Olaniyan, O.T., Adeyomoye, O., Dare, A., Adeniyi, M.J., Alex, E., Rebezov, M., Isabekova, O., Ali Shariati, M. (2022d). Smart Sensing for COVID-19 Pandemic. In: Pani, S.K., Dash, S., dos Santos, W.P., Chan Bukhari, S.A., Flammini, F. (eds), *Assessing COVID-19 and Other Pandemics and Epidemics using Computational Modelling and Data Analysis*. Springer, Cham. https://doi.org/10.1007/978-3-030-79753-9_9

Adetunji, C.O., Olaniyan, O.T., Adeyomoye, O., Dare, A., Adeniyi, M.J., Alex, E., Rebezov, M., Petukhova, E., Ali Shariati, M. (2022e). Internet of Health Things (IoHT) for COVID-19. In: Pani, S.K., Dash, S., dos Santos, W.P., Chan Bukhari, S.A., Flammini, F. (eds). *Assessing COVID-19 and Other Pandemics and Epidemics using Computational Modelling and Data Analysis*. Springer, Cham. https://doi.org/10.1007/978-3-030-79753-9_5

Adetunji, C.O., Olaniyan, O.T., Adeyomoye, O., Dare, A., Adeniyi, M.J., Alex, E., Rebezov, M., Koriagina, N., Ali Shariati, M. (2022f). Diverse Techniques Applied for Effective Diagnosis of COVID-19. In: Pani S.K., Dash S., dos Santos W.P., Chan Bukhari S.A., Flammini F. (eds), *Assessing COVID-19 and Other Pandemics and Epidemics using Computational Modelling and Data Analysis*. Springer, Cham. https://doi.org/10.1007/978-3-030-79753-9_3

Adetunji, C.O., Nwankwo, W., Olayinka, A.S., Olugbemi, O.T., Akram, M., Laila, U., Olugbenga, M.S., Oshinjo, A.M., Adetunji, J.B., Okotie, G.E,, & (Diuto) Esiobu, N. (2022g). Machine Learning and Behaviour Modification for COVID-19.DOI: 10.1201/9781003178903-17. In: *Medical Biotechnology, Biopharmaceutics, Forensic Science and Bioinformatics*. CRC Press, p. 17. eBook ISBN 9781003178903.

Adetunji, C. O., Olugbemi, O.T., Akram, M., Laila, U., Samuel, M.O., Oshinjo, A.M., Adetunji, J.B., Okotie, G.E., (Diuto) Esiobu, N., Oyedara, O.O., & Adeyemi, F.M. (2022h). Application of Computational and Bioinformatics Techniques in Drug Repurposing for Effective Development of Potential Drug Candidate for the Management of COVID-19. In: *Medical Biotechnology, Biopharmaceutics, Forensic Science and Bioinformatics*. CRC Press, p.14. eBook ISBN 9781003178903. DOI: 10.1201/9781003178903-15

Adetunji, C.O. & Oyeyemi, O.T. (2022). Antiprotozoal Activity of Some Medicinal Plants against Entamoeba Histolytica, the Causative Agent of Amoebiasis. In: *Medical Biotechnology, Biopharmaceutics, Forensic Science and Bioinformatics*. CRC Press, p. 12. eBook ISBN 9781003178903. www.taylorfrancis.com/chapters/edit/10.1201/9781003178903-20/antiprotozoal-activity-medicinal-plants-entamoeba-histolytica-causative-agent-amoebiasis-charles-oluwaseun-adetunji-oyetunde-oyeyemi.

Adetunji, C.O. & Ukhurebor, K.E. (2021). Recent Trends in Utilization of Biotechnological Tools for Environmental Sustainability. In: Adetunji, C.O., Panpatte, D.G., Jhala, Y.K. (eds), *Microbial Rejuvenation of Polluted Environment. Microorganisms for Sustainability*, vol. 27. Springer, Singapore. https://doi.org/10.1007/978-981-15-7459-7_11

Adetunji, C.O., Fawole, O.B., Afolayan, S.S., Olaleye, O.O., & Adetunji, J.B. (2012). An Oral Pesentation during 3rd NISFT Western Chapter Half Year Conference/General Meeting, Ilorin, pp. 14–16.

Adley C. C. (2014). Past, Present and Future of Sensors in Food Production. *Foods (Basel, Switzerland)*, *3*(3), 491–510.

Akram, M., Adetunji, C.O., Egbuna, C., Jabeen, S., Olaniyan, O., Ezeofor, N.J., Anani, O.A., Laila, U., Găman, M.-A., Patrick-Iwuanyanwu, K., Ifemeje, J.C., Chikwendu, C.J., Michael, O.C., & Rudrapal, M. (2021a). Dengue Fever, Chapter 17. In: *Neglected Tropical Diseases and Phytochemicals in Drug Discovery*. Wiley. DOI: 10.1002/9781119617143

Akram, M., Mohiuddin, E., Adetunji, C.O., Oladosun, T.O., Ozolua, P., Olisaka, F.N., Egbuna, C., Michael, O., Adetunji, J.B., Hameed, L., Awuchi, C.G., Patrick-Iwuanyanwu, K., & Olaniyan, O. (2021b). Prospects of Phytochemicals for the Treatment of Helminthiasis. In: *Neglected Tropical Diseases and Phytochemicals in Drug Discovery*. Chapter 7: Wiley. DOI: 10.1002/9781119617143

Arowora, K. A., Abiodun, A.A., Adetunji, C. O., Sanu, F. T., Afolayan, S. S., & Ogundele, B. A. (2012). Levels of Aflatoxins in Some Agricultural Commodities Sold at Baboko Market in Ilorin, Nigeria. *Global Journal of Science Frontier Research, 12*(10), 31–33.

Banús, N., Boada, I., Xiberta, P., Toldrà, P., & Bustins, N. (2021). Deep Learning for the Quality Control of Thermoforming Food Packages. *Scientific Reports*, *11*(1), 21887.

Barnes, J., Whiley, H., Ross, K., & Smith, J. (2022). Defining Food Safety Inspection. *International Journal of Environmental Research and Public Health, 19*(2), 789.

Chapman, B., & Gunter, C. (2018). Local Food Systems Food Safety Concerns. *Microbiology Spectrum*, *6*(2), 10.1128/microbiolspec.PFS-0020-2017

Cheng, M. S., & Toh, C. S. (2013). Novel Biosensing Methodologies for Ultrasensitive Detection of Viruses. *The Analyst*, *138*(21), 6219–6229.

Choi, J. R., Yong, K. W., Choi, J. Y., & Cowie, A. C. (2019). Emerging Point-of-care Technologies for Food Safety Analysis. *Sensors (Basel, Switzerland)*, *19*(4), 817.

Connelly, J. T., & Baeumner, A. J. (2012). Biosensors for the Detection of Waterborne Pathogens. *Analytical and Bioanalytical Chemistry*, *402*(1), 117–127.

Dauda, W.P., Abraham, P., Glen, E., Adetunji, C.O., Ghazanfar, S., Ali, S., Al-Zahrani, M., Azameti, M.K., Alao, S.E.L., Zarafi, A.B., Abraham, M.P., & Musa, H. (2022a). Robust Profiling of Cytochrome P450s (P450ome) in Notable *Aspergillus spp*. *Life*, *12*(3), 451. https://doi.org/10.3390/life12030451

Dauda, W.P., Morumda, D., Abraham, P., Adetunji, C.O., Ghazanfar, S., Glen, E., Abraham, S.E., Peter, G.W., Ogra, I.O., Ifeanyi, U.J., Musa, H., Azameti, M.K., Paray, B.A., & Gulnaz, A. (2022b). Genome-wide Analysis of Cytochrome P450s of Alternaria Species: Evolutionary Origin, Family Expansion and Putative Functions. *Journal of Fungi, 8*(4), 324. https://doi.org/10.3390/jof8040324

Farber, J. M., & Peterkin, P. I. (1991). *Listeria Monocytogenes*, A Food-borne Pathogen. *Microbiological Reviews*, *55*(3), 476–511.

Fu, Y., & Dudley, E. G. (2021). Antimicrobial-coated Films as Food Packaging: A Review. *Comprehensive Reviews in Food Science and Food Safety*, *20*(4), 3404–3437.

Gallo, M., Ferrara, L., Calogero, A., Montesano, D., & Naviglio, D. (2020). Relationships Between Food and Diseases: What to Know to Ensure Food Safety. *Food Research International (Ottawa)*, *137*, 109414.

Geeraerd, A.H., Herremans, C.H., Cenens, C., & Van Impe, J.F. (1998). Application of Artificial Neural Networks as a Non-linear Modular Modeling Technique to Describe Bacterial Growth in Chilled Food Products. *International Journal of Food Microbiology*, *44*(1–2), 49–68.

Gould, L.H., Walsh, K.A., Vieira, A.R., Herman, K., Williams, I.T., Hall, A.J., Cole, D., & Centers for Disease Control and Prevention (2013). Surveillance for Foodborne Disease Outbreaks – United States, 1998–2008. *Morbidity and Mortality Weekly Report. Surveillance Summaries (Washington, DC: 2002)*, *62*(2), 1–34.

Gumienna, M., & Górna, B. (2021). Antimicrobial Food Packaging with Biodegradable Polymers and Bacteriocins. *Molecules (Basel, Switzerland)*, *26*(12), 3735.

Huang, Y., Kangas, L.J., & Rasco, B.A. (2007). Applications of Artificial Neural Networks (ANNs) in Food Science. *Critical Reviews in Food Science and Nutrition*, *47*(2), 113–126.

Jiang, J., Zhang, M., Bhandari, B., & Cao, P. (2020). Current Processing and Packing Technology for Space Foods: A Review. *Critical Reviews in Food Science and Nutrition*, *60*(21), 3573–3588.

Jiang, X., Li, D., Xu, X., Ying, Y., Li, Y., Ye, Z., & Wang, J. (2008). Immunosensors for Detection of Pesticide Residues. *Biosensors & Bioelectronics*, *23*(11), 1577–1587.

Jin, X., Liu, C., Xu, T., Su, L., & Zhang, X. (2020). Artificial Intelligence Biosensors: Challenges and Prospects. *Biosensors & Bioelectronics*, *165*, 112412.

Kadariya, J., Smith, T.C., & Thapaliya, D. (2014). *Staphylococcus aureus* and Staphylococcal Food-borne Disease: An Ongoing Challenge in Public Health. *BioMed Research International*, *2014*, 827965.

Kraśniewska, K., Galus, S., & Gniewosz, M. (2020). Biopolymers-Based Materials Containing Silver Nanoparticles as Active Packaging for Food Applications-A Review. *International Journal of Molecular Sciences*, *21*(3), 698.

Lahiri, D., Nag, M., Sarkar, T., Dutta, B., & Ray, R. R. (2021). Antibiofilm Activity of α-Amylase from *Bacillus subtilis* and Prediction of the Optimized Conditions for Biofilm Removal by Response Surface Methodology (RSM) and Artificial Neural Network (ANN). *Applied Biochemistry and Biotechnology*, *193*(6), 1853–1872.

Li, Z., & Sheng, C. (2014). Nanosensors for Food Safety. *Journal of Nanoscience and Nanotechnology*, *14*(1), 905–912.

Liu, B., Gurr, P.A., & Qiao, G. G. (2020). Irreversible Spoilage Sensors for Protein-Based Food. *ACS Sensors*, *5*(9), 2903–2908.

Lou, W., & Nakai, S. (2001). Artificial Neural Network-based Predictive Model for Bacterial Growth in a Simulated Medium of Modified-atmosphere-packed Cooked Meat Products. *Journal of Agricultural and Food Chemistry*, *49*(4), 1799–1804.

Mao, Z., Chen, R., Wang, X., Zhou, Z., Peng, Y., Li, S., Han, D., Li, S., Wang, Y., Han, T., Liang, J., Ren, S., & Gao, Z. (2022). CRISPR/Cas12a-based Technology: A Powerful Tool for Biosensing in Food Safety. *Trends in Food Science & Technology*, *122*, 211–222.

Mijanur Rahman, A., Kim, D. H., Jang, H. D., Yang, J. H., & Lee, S. J. (2018). Preliminary Study on Biosensor-Type Time-Temperature Integrator for Intelligent Food Packaging. *Sensors (Basel, Switzerland)*, *18*(6), 1949.

Neethirajan, S., Ragavan, V., Weng, X., & Chand, R. (2018). Biosensors for Sustainable Food Engineering: Challenges and Perspectives. *Biosensors*, *8*(1), 23.

Nešić, A., Cabrera-Barjas, G., Dimitrijević-Branković, S., Davidović, S., Radovanović, N., & Delattre, C. (2019). Prospect of Polysaccharide-Based Materials as Advanced Food Packaging. *Molecules (Basel, Switzerland)*, *25*(1), 135.

Nwankwo, W., Adetunji, C.O., Ukhurebor, K.E., Panpatte, D.G., Makinde, A.S., & Hefft, D.I. (2021). Recent Advances in Application of Microbial Enzymes for Biodegradation of Waste and Hazardous Waste Material. In: Adetunji, C.O., Panpatte, D.G., & Jhala, Y.K. (eds), *Microbial Rejuvenation of Polluted Environment. Microorganisms for Sustainability*, vol. 27. Springer, Singapore. https://doi.org/10.1007/978-981-15-7459-7_3

Nyachuba D.G. (2010). Foodborne Illness: Is It on the Rise? *Nutrition Reviews*, *68*(5), 257–269.

Okeke, N.E., Adetunji, C.O., Nwankwo, W., Ukhurebor, K.E., Makinde, A.S., & Panpatte, D.G. (2021). A Critical Review of Microbial Transport in Effluent Waste and Sewage Sludge Treatment. In: Adetunji, C.O., Panpatte, D.G., Jhala, Y.K. (eds), *Microbial Rejuvenation of Polluted Environment. Microorganisms for Sustainability*, vol. 27. Springer, Singapore. https://doi.org/10.1007/978-981-15-7459-7_10

Olaniyan, O.T., Adetunji, C.O., Adeniyi, M.J., & Hefft, D.I. (2022a). Machine Learning Techniques for High-Performance Computing for IoT Applications in Healthcare. In: *Deep Learning, Machine Learning and IoT in Biomedical and Health Informatics*. CRC Press, p. 13. eBook ISBN 9780367548445. DOI: 10.1201/9780367548445-20

Olaniyan, O.T., Adetunji, C.O., Adeniyi, M.J., & Hefft, D.I. (2022b). Computational Intelligence in IoT Healthcare. In: *Deep Learning, Machine Learning and IoT in Biomedical and Health Informatics*. CRC Press, p. 13. eBook ISBN 9780367548445. DOI: 10.1201/9780367548445-19

Oluwaseun, A.C., Phazang, P., & Sarin, N.B. (2017). Significance of Rhamnolipids as a Biological Control Agent in the Management of Crops/Plant Pathogens. *Current Trends in Biomedical Engineering & Biosciences, 10* (3), 54–55. Juniper Publishers Inc.

Oyedara, O.O., Adeyemi, F.M., Adetunji, C.O., Elufisan, T.O. (2022). Repositioning Antiviral Drugs as a Rapid and Cost-Effective Approach to Discover Treatment against SARS-CoV-2 Infection. In: *Medical Biotechnology, Biopharmaceutics, Forensic Science and Bioinformatics*. CRC Press, p. 12. eBook ISBN 9781003178903. DOI: 10.1201/9781003178903-10.

Palanichamy, A., Jayas, D. S., & Holley, R. A. (2008). Predicting Survival of *Escherichia coli* O157:H7 in Dry Fermented Sausage Using Artificial Neural Networks. *Journal of Food Protection*, *71*(1), 6–12.

Palchetti, I., Laschi, S., & Mascini, M. (2009). Electrochemical Biosensor Technology: Application to Pesticide Detection. *Methods in Molecular Biology (Clifton, NJ)*, *504*, 115–126.

Patnode, K., Rasulev, B., & Voronov, A. (2022). Synergistic Behavior of Plant Proteins and Biobased Latexes in Bioplastic Food Packaging Materials: Experimental and Machine Learning Study. *ACS Applied Materials & Interfaces*, *14*(6), 8384–8393.

Pejcic, B., De Marco, R., & Parkinson, G. (2006). The Role of Biosensors in the Detection of Emerging Infectious Diseases. *The Analyst*, *131*(10), 1079–1090.

Petersen, M., Yu, Z., & Lu, X. (2021). Application of Raman Spectroscopic Methods in Food Safety: A Review. *Biosensors*, *11*(6), 187.

Rodrigues, C., Souza, V., Coelhoso, I., & Fernando, A. L. (2021). Bio-Based Sensors for Smart Food Packaging-Current Applications and Future Trends. *Sensors (Basel, Switzerland)*, *21*(6), 2148.

Rogers, J. K., Taylor, N. D., & Church, G. M. (2016). Biosensor-based Engineering of Biosynthetic Pathways. *Current Opinion in Biotechnology*, *42*, 84–91.

Sangeetha, J., Hospet, R., Thangadurai, D., Adetunji, C.O., Islam, S., Pujari, N., & Al-Tawaha, A.R.M.S. (2021). Nanopesticides, Nanoherbicides, and Nanofertilizers: The Greener Aspects of Agrochemical Synthesis Using Nanotools and Nanoprocesses Toward Sustainable Agriculture. In: Kharissova, O.V., Torres-Martínez, L.M., Kharisov, B.I. (eds), *Handbook of Nanomaterials and Nanocomposites for Energy and Environmental Applications*. Springer, Cham. https://doi.org/10.1007/978-3-030-36268-3_44

Sun, Q., Zhang, M., & Mujumdar, A. S. (2019). Recent Developments of Artificial Intelligence in Drying of Fresh Food: A Review. *Critical Reviews in Food Science and Nutrition*, *59*(14), 2258–2275.

Tittl, A., John-Herpin, A., Leitis, A., Arvelo, E. R., & Altug, H. (2019). Metasurface-Based Molecular Biosensing Aided by Artificial Intelligence. *Angewandte Chemie (International ed. in English)*, *58*(42), 14810–14822.

Todd, E. C. (1997). Epidemiology of Foodborne Diseases: A Worldwide Review. *World Health Statistics Quarterly. Rapport trimestriel de statistiques sanitaires mondiales*, *50*(1–2), 30–50.

Torrence, M. E. (2016). Introduction to Preharvest Food Safety. *Microbiology Spectrum*, *4*(5). 10.1128/microbiolspec.PFS-0009-2015.

Trajkovska Petkoska, A., Daniloski, D., D'Cunha, N. M., Naumovski, N., & Broach, A. T. (2021). Edible Packaging: Sustainable Solutions and Novel Trends in Food Packaging. *Food Research International (Ottawa)*, *140*, 109981.

Ukhurebor, K.E., Mishra, P., Mishra, R.R., & Adetunji, C.O. (2020). Nexus Between Climate Change and Food Innovation Technology: Recent Advances. In: Mishra, P., Mishra, R.R., Adetunji, C.O. (eds), *Innovations in Food Technology*. Springer, Singapore. https://doi.org/10.1007/978-981-15-6121-4_20

Ukhurebor, K. E., Adetunji, C.O., Bobadoye, A.O., Aigbe, U.O., Onyancha, R.B., Siloko, I.U., Emegha, J.O., Okocha, G.O., & Abiodun, I.C. (2021a). Bionanomaterials for Biosensor Technology. *Bionanomaterials. Fundamentals and Biomedical Applications*, 5–22.

Ukhurebor, K.E., Nwankwo, W., Adetunji, C.O., & Makinde, A.S. (2021b). Artificial Intelligence and Internet of Things in Instrumentation and Control in Waste Biodegradation Plants: Recent Developments. In: Adetunji, C.O., Panpatte, D.G., Jhala, Y.K. (eds), *Microbial Rejuvenation of Polluted Environment. Microorganisms for Sustainability*, vol 27. Springer, Singapore. https://doi.org/10.1007/978-981-15-7459-7_12

Wang, K., Chen, L., Ma, X., Ma, L., Chou, K. C., Cao, Y., Khan, I., Gölz, G., & Lu, X. (2020). *Arcobacter* Identification and Species Determination Using Raman Spectroscopy Combined with Neural Networks. *Applied and Environmental Microbiology*, *86*(20), e00924–20.

Wendlandt, S., Schwarz, S., & Silley, P. (2013). Methicillin-resistant *Staphylococcus aureus*: A Food-borne Pathogen? *Annual Review of Food Science and Technology*, *4*, 117–139.

3 Optical Methods

Advanced Image Recognition and Feature Extraction Solutions for Food Manufacturing

Olugbemi T. Olaniyan, Charles Oluwaseun Adetunji, Olorunsola Adeyomoye, Akinola Samson Olayinka, and Olalekan Akinbo

CONTENTS

3.1 INTRODUCTION

Studies have projected that there is going to be population explosion by the year 2050, placing a huge demand on food, thus food industries must increase food production and quality to meet the growing demands. In recent years great efforts have been channeled towards improvement in food production and in quality assessment using automated food classification utilizing specific features. Different modules are utilized in the classification of grains, modules such as image acquisition, segmentation, preprocessing, classification. In image acquisition, such as visible, hyperspectral or multispectral based approaches are utilized to obtain information. Image processing involves cropping, color space mapping, noise filtering, sharpness enhancement, color conversion, contrast, and scaling that will overall enhance the quality of the image. Many low-quality images will pose as a challenge to the image classification process – such as noisy data, poor lighting, low contrast effect, and objects. The process of segmentation allows an image to be sectioned into different regions using thresholding and classification pattern recognition, like support vector machine, artificial neural network, linear discriminant, k-means clustering, k-Nearest Neighbor analysis.

Food is a vital necessity providing warmth, growth, and repair to worn out tissues in human body. There are diverse food varieties such as fruit, nuts, vegetables, seeds,

DOI: 10.1201/9781003207955-3

fish, meat, and eggs. These raw food varieties are converted to different products like bread, milk, butter and biscuits through grinding, and cooking, and are packaged for consumption. Various packaging modes like bottles, boxes, cans, packs, packets are available. The quality of food products requires special consideration to guarantee standards during production to minimize contamination. Technological advancement has led to intelligent packaging and food technology for quality control, production, monitoring, distribution, and storage.

Computer visions are flexible automation for the analysis of digital images and construct computational image processing for the performance of tasks. Some of the machine vision tasks in food industry are picking, pruning, spraying, sizing, evisceration, icing pattern, inspection and monitoring, and packing. The quality and safety of food products are very important, thus ensuring the development of machine vision in all aspects of food processing – such as cultivation, storage, harvesting, transportation, and packaging – will increase safety and quality measures. These machine visions will promote food grading, removing impurities, detection of pathogenic organisms, crop monitoring, and improving efficiency.

Recent vision systems are based on sensors that constitute computer vision and technologies such as hyperspectral and multispectral cameras, new sensors, terahertz and biospeckle cameras, and novel artificial intelligence techniques. These sensors can be applied to detect nondestructive techniques in determining mechanical properties, quality parameters, composition, identification of defects, appearance, vegetables and fruits grading, plant disease detection for smart farming, three-dimensional (3D) reconstruction, and advanced quality control of fruit post-harvest. Deep learning approach is also utilized in dietary appraisal for the treatment of weight loss in modern computer-based food recognition system (Adetunji et al., 2020; Adetunji et al., 2020a,b,c; Oyedara et al., 2022; Adetunji et al., 2022a,cb;d,e,f,g,h; Adetunji and Oyeyemi, 2022i; Olaniyan et al., 2022). This chapter reviewed recent articles on computer vision system application in agricultural and food industry for food processing and classification using image processing techniques. Figure 3.1 shows the types of cameras that are applied in computer vision and artificial intelligence, while Figure 3.2 shows the various areas of sensors application.

3.2 ADVANCED IMAGE RECOGNITION AND FEATURE EXTRACTION SOLUTIONS FOR FOOD MANUFACTURING

Patel et al. (2015) reported that the utilization of computer and image processing technology to maintain food safety and quality has gained tremendous interest in food science. The authors noted that the process of imaging involves noise removal, mid-level image contrast enhancing, segmentation based on thresholding, region, classification, and gradient methods. In the last few decades, artificial neural networks have been utilized in the classification, prediction, analysis and modeling of food safety and quality. Food particles can be classified based on structural appearance, texture, histogram, and flavor, which are external factors, and microbial, chemical and physical factors, which are internal factors.

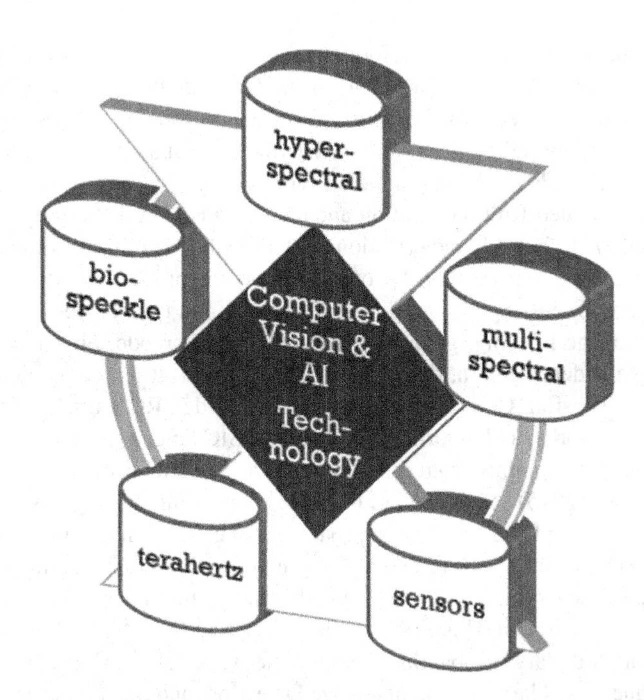

FIGURE 3.1 Camera Types in Computer Vision and AI.

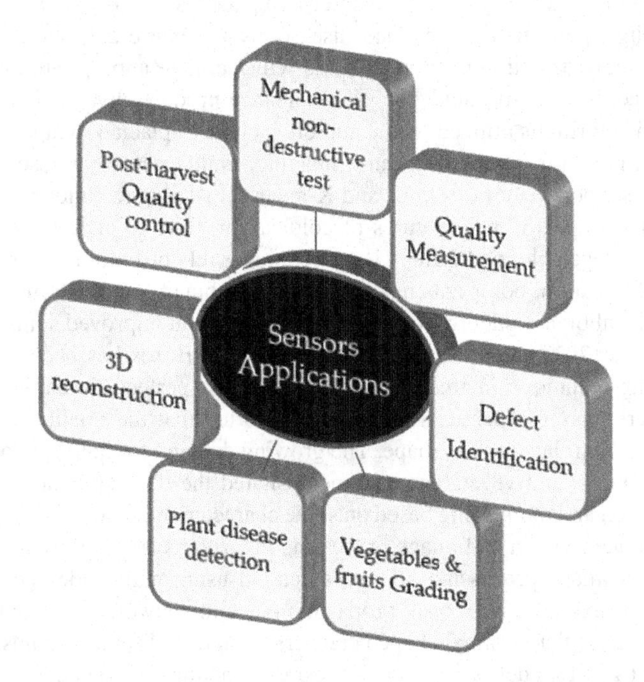

FIGURE 3.2 Various Areas of Sensors Application.

Henry et al. (2021) reported that food grain classification can be done using computer vision techniques which are state of art facilities that can classify different grains. The authors noted that different images are considered in these approaches, such as images from multispectral, infrared, hyperspectral and visible bands.

Jyoti and Balaji (2012) analyzed the use of computer vision systems and imaging analysis in automated fruit recognition and classification. In their study, the authors demonstrated that fruit characterization and classification have suffered from a number of challenges that must be overcome before proper classification can be applied. Image processing and computer algorithms are utilized for quality control measures, package labeling, geometric measurement, barcode, object sorting, fault inspection, part identification, defect location and localization, contaminations, and insect or pest invasion. Computer vision like HSI model, RGB model, Fuzzy logic, MLP-neural networks, and algorithm in Visual BASIC language are being utilized to extract relevant information from different fruits and vegetables.

Drashti et al. (2015) reported that, over the years, fruit grading has moved from the traditional method to automated approaches using different computer algorithms. In computer algorithms, texture and shape are very important imaging features for digital image processing. The authors noted some of the techniques in image processing, such as Circular Hough Transform, Discrete Wavelet Transform, Fourier Transform of Boundary, Probabilistic Neural Network, Chain Code, fractal dimension technique, Boundary Tracing and Edge Detection, and Scale Space.

Shiv and Anand (2015) revealed that in the agricultural sector images are utilized for information generation and data sources. Vegetables and fruits are classified to detect quality, appropriate prizes, and disease, using texture and color segregation methods. Timely detection of crop diseases will facilitate appropriate management through insecticides or pesticides so as to enhance productivity. Some of the computer vision algorithms utilized by the authors include Laplacian pyramid transform, fuzzy logic, near-infrared hyperspectral imaging, multispectral fluorescence, multiclass kernel support vector machine, and K-means clustering technique. Many of the extraction techniques utilize features of color and texture to efficiently and accurately validate vegetables and fruits. These techniques rely on color histogram, border/interior classification, color coherence vector, local binary pattern, Unser's feature, completed local binary patterns, difference histogram and improved sum.

Maya et al. (2017) demonstrated that Matlab Tool Boxes has been shown to be the emerging technique in imaging processing for a perfect simulation platform, which is very easy to operate. Ding (1994) reported that the quality of food substance is directly related to the shape. The growing demand for quality food products like fruits, grains, and vegetables have necessitated the development of automated food inspection and monitoring based on some characteristic features using computer vision. Computer vision and image processing algorithms are very accurate and fast techniques in image processing and classification using multi-index active model-based feature extractor and back-propagation neural network, minimum indeterminate zone classifiers, animal-shaped crackers, corn kernels and almonds. There are so many statistical model-based variant feature extraction methods that are deployed for inspection of shape and texture.

Jia et al. (2018) revealed that chronic diseases like hypertension, cancer, cardiovascular disorders, obesity, and diabetes including some neglected tropical diseases are related or associated with dietary-linked factors such as poor lifestyle and unhealthy diet (Akram et al., 2021a,b). The authors revealed that the traditional methods of dietary monitoring have many deficiencies, such as imprecision, time consumption, low adherence, and underreporting. Recent advancement in technologies have revealed that computer vision algorithms can be employed as food monitoring platforms, such as K-Nearest Neighbor, Convolutional Neural Network, Ensemble Net, Deep Convolutional Neural Network, and Inception-v3, as automated assessment, volume estimation, food recognition, like machine learning. The pixel binning technique has attracted much interest in spectroscopy and image acquisition. This technique has the ability to combine larger pixels, reduce noise impact, and achieve accurate classification with enhanced sensitivity.

Nidhi et al. (2022) reported that in food technology, utilization of artificial intelligence has increased the production level so as to meet the growing demand on food. The artificial intelligence is utilized in the determination of food quality, classification of food, control tools, food safety, pathogenic organism determination, and prediction. There are so many platforms and algorithms such as fuzzy logic, adaptive neuro-fuzzy inference system, ANN, random forest regression, radial basis function networks, multilayer perceptron, stepwise linear regression, ordinary least square regression, principal component regression, partial least square regression, support vector regression, k-nearest neighbors' regression, boosted logistic regression and machine learning modeling that are deployed in food industries.

Yi et al. (2019a) reported that deep learning-based computer vision systems like convolutional neural networks is very important data sensing application, but the limitations are quite enormous, like non-energy-saving features making it difficult for many mobile systems. The authors reported that convolutional neural networks are widely applied to several image functions such as image recognition, image resolution, nutrition tracking, image augmentation, and food classification utilizing Xception (Yi et al. 2019b).

Indrajeet et al. (2021) reported that in the entire world, food industries remained the most significant manufacturing industry with the largest amount of employability. Due to population explosion and environmental contaminations, food safety and quality remains the biggest challenge, thus industrial food automation use seems to the best approach to resolve these food issues. Food automation is based on deep learning, machine learning and artificial intelligence algorithms to enhance food production, safe delivery, quality and competence.

Rijwan et al. (2021) revealed that in maintaining food safety, certain factors like transporting, preparation, and storage must be hygienic and are very crucial in the prevention of foodborne pathogens. The authors noted that computer visions are available to facilitate automatic, economic and nondestructive approaches in the monitoring of food safety. Many vegetables and fruits are classified and assessed utilizing image processing techniques. Food substances have different forms, quality, flavor, thickness, texture and contents – thus making selection or assessment to be very difficult, so identification can be done through deep learning techniques like convolutional neural network and image processing like image segmentation, defect segmentation, and feature extraction involving reduction in the total dimension in the raw data to achieve grouping processing using Global Color Histogram, Color Coherence Vector, Border/Interior Classification, Local Binary Pattern, Completed Local Binary Pattern, Unser's Feature, and Improved Sum and Difference Histogram.

Qiu (2016) revealed that food sfety is an important topic to the public, with serious health implications if not addressed appropriately as more food contaminations are being witnessed in recent years. These food contaminants can be traced to the recent increase in population and growing demand on the limited available food. Many scientists have traced different contaminants like insecticides, herbicides, and pesticides in food samples to various health diseases and sickness in humans. The past traditional method of extraction such as Soxhlet extraction and liquid-liquid extraction are cumbersome, expensive, and time wasting. Recent advancement in research has provided better means of extraction, such as solid phase extraction and accelerated solvent extraction. They provide faster, less consumable, reduced pollution, and less expensive methods.

Iniran et al. (1990) reported that advancement in optical technologies have generated interest among biomedical scientists due to their important characteristics, such as non-interference, high speed, and massive parallelism. These technologies are utilized for image processing, pattern recognition, signal processing, and matrix vector multiplication. Some of the approaches include spatial filtering of coherent light, matched filters, acoustooptic devices, discrete optical processors, spatial-temporal integration, and symbolic substitution in computer vision.

Suganya (2015) reported that, in image processing, feature extraction techniques constitute an important methods that are applied, such as thresholding, binarization, normalization and resizing. The main usefulness of this technique is to extract important characteristics, patterns or information for image recognition and classification. Mobile cameras have produced more food pictures in terms of the type of food eaten by some individuals, which are referred to as automated food journaling.

Valous and Sun (2012) revealed that image processing and analysis involves automated image capturing for measurement, classification, extraction, and segmentation. Recently, these techniques are utilized for monitoring food, for beverages and fruits production, and safety and quality, which are produced in large scale. It is known that safety and quality of food products are important issues to society in order to maintain good health. In food production, like cultivation, storage, harvesting, packaging, transporting, and consumption, which involves intense labor and food contact that may trigger contamination, food grading, spoilage and pathogenic invasion. Computer vision has been reported to improve efficiency and quality of food processing through image processing utilizing deep learning and machine-learning algorithms.

Anran and Carol (2021) noted that artificial intelligence, like deep learning deep neural networks, has generated tremendous interest in image processing for classification, and recognition in optical coherence tomography, and optical coherence tomography angiography for disease detection, image quality control, functional change, structural segmentation and prognosis prediction. To achieve excellent results in image processing when utilizing deep learning models, some impediment much be put into consideration like sample size, model robustness, data preprocessing standardization, performance cross-validation, and results explanation. Extraction techniques for food and other wide ranges of agricultural products are maceration like (infusion, digestion, decoction, percolation, hot continuous extraction, fermentation, counter-current extraction, sonication, and supercritical fluid extraction).

Pushpavalli (2019) highlighted the importance of computer vision techniques in fruit grading. It is reported that manual fruit grading by humans is grossly inefficient, laborious and full of errors. Thus, the need for automated fruit grading systems like neural networks, and support vector machines using color, texture, weight, and size to classify, identify and sort quality products.

Loke et al. (2018) revealed that much of the information generated from the computer vision is converted to readable machine formats using software optical mark recognition. Computer-aided image processing systems utilize food features like shape and size based recognition and classification of microbial safety, defects detection, variety determination, sorting and quality grading in food industry.

Predrag et al. (2018) reported that biological active agents in foods can be extracted using microwave assisted extraction high-pressure assisted extraction, ultrasound-assisted extraction, pulsed electric fields assisted extraction, high voltage electric discharges assisted extraction and supercritical fluids extraction.

Vijay et al. (2020) reported that emerging development or advancement in computer applications and algorithms such as artificial intelligence has generated huge data for various roles such as prediction, classification, identification, and recognition. The technologies have transformed the food industry and contributed significantly through monitoring of food processing, detection of pathogenic organisms, active packaging, and storage to facilitate sustainable food production. Park et al. (2019) revealed that food recognition and image detection in Korean food for the estimation of dietary intake using mobile devices have generated huge interest among biomedical scientists. The authors demonstrated that, from their study, K-foodNet which is a Deep Convolutional Neural Network, produced the most accurate and rapid results. Dietary assessment is important to monitor personal food intake, conduct nutritional research, public health safety, management of chronic diseases, and reduce health burden.

Juliana et al. (2020) noted that human activity, particularly in agricultural sector has witnessed automation in recent times. This can be attributed to the advancement in computer applications, platforms, and algorithms for pest detection, image recognition, identification, and classification. These technologies are key factors for a global food industrial revolution so as to meet the increasing demand in food industry due to a population explosion. Computer vision systems cut across every production stage of food processing, such as planting, farm management, cultivation, disease control, robotic harvesting, weed control, transportation, distribution, and storage. Computer vision hyperspectral imaging systems can obtain information and huge data, not visible to the naked human eye, for the purpose of analysis, detection, classification, monitoring, identification, selection, and processing. Unmanned aerial vehicle, Internet of Things, robust algorithms, deep learning, grey level co-occurrence matrix,

InceptionV3, MobileNet model, fully convolutional networks, surface enhanced Raman spectroscopy, naive Bayesian, k-nearest neighbor, linear discriminant analysis, and decision tree are the various computer vision systems that can be utilized to process images derived from food products based on size, shape, color, weight, and texture.

Rohit et al. (2021) revealed that color, size, shape, texture and geometric characteristics are the physical features in wheat for refraction and classification. The authors were able to analyze the wheat variants using Open CV library and python programming language from deep and machine learning algorithms.

3.3 CONCLUSION

This chapter focused on the application of automated image recognition in food industry through the utilization of techniques and algorithms involved in a computer vision system. These computer vision systems provide complex operations for basic recognition of objects, patterns and features detection through the use of color, texture, shape and size. Image processing involves conversion of one form of image to another by the operation of pixels such as inverting image, copying images, stretching image, intensities, and thresholding. So many algorithms and computer programming languages are utilized in the food industry, such as artificial neural networks, Convolution Neural Network and digital image technologies, which generate certain outcomes based on the user's code. These imaging technologies are utilized to perform image enhancement, analysis, segmentation, diagnosis, and classification in the agricultural, environment and food industries (Adetunji et al., 2021a,b,c; Ukhurebor et al., 2021; Sangeetha et al., 2021; Adetunji and Ukhurebor 2021; Oluwaseun et al., 2017; Adetunji et al., 2012; Arowora et al., 2012; Dauda et al., 2022a,b; Ukhurebor et al. 2021; Okeke et al., 2021; Adetunji and Anani, 2021; Nwankwo et al., 2021; Adejumo and Adetunji. 2018; Ukhurebor et al., 2021).

REFERENCES

Adejumo, I.O., and Adetunji, C.O. (2018). Production and Evaluation of Biodegraded Feather Meal using Immobilised and Crude Enzyme from *Bacillus subtilis* on Broiler Chickens. *Brazilian Journal of Biological Sciences*, *5*(10), 405–416.

Adetunji, C.O. and Anani, O.A. (2021). Bioaugmentation: A Powerful Biotechnological Techniques for Sustainable Ecorestoration of Soil and Groundwater Contaminants. In: Panpatte, D.G., Jhala, Y.K. (eds), *Microbial Rejuvenation of Polluted Environment. Microorganisms for Sustainability*, vol. 25. Springer, Singapore. https://doi.org/10.1007/978-981-15-7447-4_15

Adetunji, C.O., Anani, O.A., Olaniyan, O.T., Inobeme, A., Olisaka, F.N., Uwadiae, E.O., and Obayagbona, O.N. (2021a). Recent Trends in Organic Farming. In: Soni, R., Suyal, D.C., Bhargava, P., Goel, R. (eds), *Microbiological Activity for Soil and Plant Health Management*. Springer, Singapore. https://doi.org/10.1007/978-981-16-2922-8_20

Adetunji, C.O., Egbuna, C., Oladosun, T.O., Akram, M., Michael, O., Olisaka, F.N., Ozolua, P., Adetunji, J.B., Enoyoze, G.E., and Olaniyan, O. (2021b). Efficacy of Phytochemicals of Medicinal Plants for the Treatment of Human Echinococcosis. Ch 8. In: *Neglected*

Tropical Diseases and Phytochemicals in Drug Discovery. Wiley. DOI: 10.1002/9781119617143

Adetunji, C.O., Fawole, O.B., Afolayan, S.S., Olaleye, O.O., and Adetunji, J.B. (2012). An Oral Presentation During 3rd NISFT Western Chapter Half Year Conference/General Meeting, Ilorin, pp. 14–16.

Adetunji, C.O., Michael, O.S., Nwankwo, W., Ukhurebor, K.E., Anani, O.A., Oloke, J.K., Varma, A., Kadiri, O., Jain, A., and Adetunji, J.B.. (2021c). Quinoa, The Next Biotech Plant: Food Security and Environmental and Health Hot Spots. In: Varma, A. (eds), *Biology and Biotechnology of Quinoa*. Springer, Singapore. https://doi.org/10.1007/978-981-16-3832-9_19

Adetunji, C.O., Mitembo, W.P., Egbuna, C., and Narasimha Rao, G.M. (2020). In Silico Modeling as a Tool to Predict and Characterize Plant Toxicity. In: Andrew G. Mtewa, Chukwuebuka Egbuna, G.M., and Narasimha Rao (eds), *Poisonous Plants and Phytochemicals in Drug Discovery*. Ch. 14. Wiley Online Libraryhttps://doi.org/10.1002/9781119650034.

Adetunji, C.O., Nwankwo, W., Olayinka, A.S., Olugbemi, O.T., Akram, M., Laila, W., Samuel, M.O., Oshinjo, A.M., Adetunji, J.B., Okotie, G.E., and (Diuto) Esiobu, N. (2022a). Computational Intelligence Techniques for Combating COVID-19. In: *Medical Biotechnology, Biopharmaceutics, Forensic Science and Bioinformatics*. CRC Press, p. 12. eBook ISBN 9781003178903. DOI: 10.1201/9781003178903-16

Adetunji, C.O., Nwankwo, W., Olayinka, A.S., Olugbemi, O.T., Akram, M., Laila, U., Olugbenga, M.S., Oshinjo, A.M., Adetunji, J.B., Okotie, G.E,, and (Diuto) Esiobu, N. (2022b). Machine Learning and Behaviour Modification for COVID-19.DOI: 10.1201/9781003178903-17. In: *Medical Biotechnology, Biopharmaceutics, Forensic Science and Bioinformatics*. CRC Press, p. 17. eBook ISBN 9781003178903.

Adetunji, C.O., Olaniyan, O.T., Adeyomoye, O., Dare, A., Adeniyi, M.J., Alex, E., Rebezov, M., Garipova, L., and Ali Shariati, M. (2022c). eHealth, mHealth, and Telemedicine for COVID-19 Pandemic. In: Pani, S.K., Dash, S., dos Santos, W.P., Chan Bukhari, S.A., & Flammini, F. (eds), *Assessing COVID-19 and Other Pandemics and Epidemics using Computational Modelling and Data Analysis*. Springer, Cham. https://doi.org/10.1007/978-3-030-79753-9_10

Adetunji, C.O., Olaniyan, O.T., Adeyomoye, O., Dare, A., Adeniyi, M.J., Alex, E., Rebezov, M., Petukhova, E., and Ali Shariati, M. (2022d). Machine Learning Approaches for COVID-19 Pandemic. In: Pani, S.K., Dash, S., dos Santos, W.P., Chan Bukhari, S.A., Flammini, F. (eds). *Assessing COVID-19 and Other Pandemics and Epidemics using Computational Modelling and Data Analysis*. Springer, Cham. https://doi.org/10.1007/978-3-030-79753-9_8

Adetunji, C. O., Olaniyan, O.T., Adeyomoye, O., Dare, A., Adeniyi, M.J., Alex, E., Rebezov, M., Isabekova, O., and Ali Shariati, M. (2022e). Smart Sensing for COVID-19 Pandemic. In: Pani, S.K., Dash, S., dos Santos, W.P., Chan Bukhari, S.A., Flammini, F. (eds), *Assessing COVID-19 and Other Pandemics and Epidemics using Computational Modelling and Data Analysis*. Springer, Cham. https://doi.org/10.1007/978-3-030-79753-9_9

Adetunji, C.O., Olaniyan, O.T., Adeyomoye, O., Dare, A., Adeniyi, M.J., Alex, E., Rebezov, M., Petukhova, E., and Ali Shariati, M. (2022f). Internet of Health Things (IoHT) for COVID-19. In: Pani, S.K., Dash, S., dos Santos, W.P., Chan Bukhari, S.A., Flammini, F. (eds). *Assessing COVID-19 and Other Pandemics and Epidemics using Computational Modelling and Data Analysis*. Springer, Cham. https://doi.org/10.1007/978-3-030-79753-9_5

Adetunji, C.O., Olaniyan, O.T., Adeyomoye, O., Dare, A., Adeniyi, M.J., Alex, E., Rebezov, M., Koriagina, N., and Ali Shariati, M. (2022g). Diverse Techniques Applied for Effective Diagnosis of COVID-19. In: Pani S.K., Dash S., dos Santos W.P., Chan Bukhari S.A., Flammini F. (eds), *Assessing COVID-19 and Other Pandemics and Epidemics using Computational Modelling and Data Analysis*. Springer, Cham. https://doi.org/10.1007/978-3-030-79753-9_3

Adetunji, C. O., Olugbemi, O.T., Akram, M., Laila, U., Samuel, M.O., Oshinjo, A.M., Adetunji, J.B., Okotie, G.E., (Diuto) Esiobu, N., Oyedara, O.O., and Adeyemi, F.M. (2022h). Application of Computational and Bioinformatics Techniques in Drug Repurposing for Effective Development of Potential Drug Candidate for the Management of COVID-19. In: *Medical Biotechnology, Biopharmaceutics, Forensic Science and Bioinformatics*. CRC Press, p.14. eBook ISBN 9781003178903. DOI: 10.1201/9781003178903-15

Adetunji, C.O. and Oyeyemi, O.T. (2022i). Antiprotozoal Activity of Some Medicinal Plants against Entamoeba Histolytica, the Causative Agent of Amoebiasis. In: *Medical Biotechnology, Biopharmaceutics, Forensic Science and Bioinformatics*. CRC Press, p. 12. eBook ISBN 9781003178903. www.taylorfrancis.com/chapters/edit/10.1201/9781003178903-20/antiprotozoal-activity-medicinal-plants-entamoeba-histolytica-causative-agent-amoebiasis-charles-oluwaseun-adetunji-oyetunde-oyeyemi.

Adetunji, C.O. and Ukhurebor, K.E. (2021). Recent Trends in Utilization of Biotechnological Tools for Environmental Sustainability. In: Adetunji, C.O., Panpatte, D.G., and Jhala, Y.K. (eds) *Microbial Rejuvenation of Polluted Environment. Microorganisms for Sustainability*, vol. 27. Springer, Singapore. https://doi.org/10.1007/978-981-15-7459-7_11

Akram, M., Adetunji, C.O., Egbuna, C., Jabeen, S., Olaniyan, O., Ezeofor, N.J., Anani, O.A., Laila, U., Găman, M.-A., Patrick-Iwuanyanwu, K., Ifemeje, J.C., Chikwendu, C.J., Michael, O.C., and Rudrapal, M. (2021a). Dengue Fever, Chapter 17. In: *Neglected Tropical Diseases and Phytochemicals in Drug Discovery*. Wiley. DOI: 10.1002/9781119617143

Akram, M., Mohiuddin, E., Adetunji, C.O., Oladosun, T.O., Ozolua, P., Olisaka, F.N., Egbuna, C., Michael, O., Adetunji, J.B., Hameed, L., Awuchi, C.G., Patrick-Iwuanyanwu, K., and Olaniyan, O. (2021b). Prospects of Phytochemicals for the Treatment of Helminthiasis. In: *Neglected Tropical Diseases and Phytochemicals in Drug Discovery*. Chapter 7: Wiley. DOI: 10.1002/9781119617143

Arowora, K. A., Abiodun, A.A., Adetunji, C. O., Sanu, F. T., Afolayan, S. S., and Ogundele, B. A. (2012). Levels of Aflatoxins in Some Agricultural Commodities Sold at Baboko Market in Ilorin, Nigeria. *Global Journal of Science Frontier Research*, *12*(10), 31–33.

Dauda, W.P., Abraham, P., Glen, E., Adetunji, C.O., Ghazanfar, S., Ali, S., Al-Zahrani, M., Azameti, M.K., Alao, S.E.L., Zarafi, A.B., Abraham, M.P., and Musa, H. (2022a). Robust Profiling of Cytochrome P450s (P450ome) in Notable *Aspergillus spp. Life*, 12(3), 451. https://doi.org/10.3390/life12030451

Dauda, W.P., Morumda, D., Abraham, P., Adetunji, C.O., Ghazanfar, S., Glen, E., Abraham, S.E., Peter, G.W., Ogra, I.O., Ifeanyi, U.J., Musa, H., Azameti, M.K., Paray, B.A., and Gulnaz, A. (2022b). Genome-wide Analysis of Cytochrome P450s of Alternaria Species: Evolutionary Origin, Family Expansion and Putative Functions. *Journal of Fungi, 8*(4), 324. https://doi.org/10.3390/jof8040324

Ding, S., and Gunasekaran, K. (1994). Shape Feature Extraction and Classification of Food Material Using Computer Vision. *American Society of Agricultural Engineers*, 37(5), 1537–1545.

Drashti, J., Patel, P., Patel, S., Ahir, B., Patel, K., and Dixit, M. (2015) Review of Shape and Texture Feature Extraction Techniques for Fruits. *International Journal of Computer Science and Information Technologies*, 6(6), 4851–4854.

Fracarolli, J.A., Pavarin, F.F.A., Castro, W., and Blasco, J. (2020). Computer Vision Applied to Food and Agricultural Products. *Revista Ciência Agronômica*, 51, *Special Agriculture* 4.0, e20207749.

Health Informatics. CRC Press, p. 13. eBook ISBN 9780367548445. DOI: 10.1201/9780367548445-20.

Iniran, G., Arif, G., and Sheikh, S.A. (1990). Optical Image Processing: The Current State of the Art. 24. ACSSC-1219010553. *IEEE* 553–556.

Jia, W., Li, Y., Qu, R., Baranowski, T., Burke, L.E., Bai, Y., Mancino, J.M., Xu, G., Mao, Z-H., and Mingui, S. (2018). Automatic food detection in egocentric images using artificial intelligence technology. *Public Health Nutrition*, 22, 1–12. 10.1017/S1368980018000538

Kakani, V., Huan Nguyen, V., Praveen Kumar, B., Kim, H., and Pasupuleti, V.R. (2020). A Critical Review on Computer Vision and Artificial Intelligence in Food Industry. *Journal of Agriculture and Food Research* 2, 100033.

Khan, R., Kumar, S., Dhingra, N., and Bhati, N. (2021). The Use of Different Image Recognition Techniques in Food Safety: A Study. *Hindawi Journal of Food Quality*. Article ID 7223164. https://doi.org/10.1155/2021/7223164.

Kodagali, J.A., and Balaji, S. (2012). Computer Vision and Image Analysis Based Techniques for Automatic Characterization of Fruits – A Review. *International Journal of Computer Applications*, (0975–8887) 50(6), 6–12.

Kumar, I., Rawat, J., Mohd, N., and Husain, S. (2021). Opportunities of Artificial Intelligence and Machine Learning in the Food Industry. *Hindawi Journal of Food Quality*, Vol. 2021, Article ID 4535567. https://doi.org/10.1155/2021/4535567.

Loke, S. C., Kasmiran K. A., and Haron S. A. (2018). A New Method of Mark Detection for Software Based Optical Mark Recognition. *PLoS One*, 13(11), e0206420. https://doi.org/10.1371/journal.pone.0206420.

Maya, V., Lakha, S.P., RajaManohar, K., Reddy, C., and Sattar, A. (2017). Advanced Image Processing Technique for Failure Analysis. *International Journal of Advanced Engineering Research and Science (IJAERS)*, 4(4): 2456–1908.

Nidhi, R.M., Mohd Ali, J., Othman, S., Hussain, M.A., Hashim, H., and Rahman, N.A. (2022). Application of Artificial Intelligence in Food Industry – A Guideline. *Food Engineering Reviews,* 14: 134–175. https://doi.org/10.1007/s12393-021-09290-z.

Nwankwo, W., Adetunji, C.O., Ukhurebor, K.E., Panpatte, D.G., Makinde, A.S., and Hefft, D.I. (2021). Recent Advances in Application of Microbial Enzymes for Biodegradation of Waste and Hazardous Waste Material. In: Adetunji, C.O., Panpatte, D.G., Jhala, Y.K. (eds), *Microbial Rejuvenation of Polluted Environment. Microorganisms for Sustainability*, vol 27. Springer, Singapore. https://doi.org/10.1007/978-981-15-7459-7_3

Okeke, N.E., Adetunji, C.O., Nwankwo, W., Ukhurebor, K.E., Makinde, A.S., Panpatte, D.G. (2021). A Critical Review of Microbial Transport in Effluent Waste and Sewage Sludge Treatment. In: Adetunji, C.O., Panpatte, D.G., Jhala, Y.K. (eds) *Microbial Rejuvenation of Polluted Environment. Microorganisms for Sustainability*, vol. 27. Springer, Singapore. https://doi.org/10.1007/978-981-15-7459-7_10

Olaniyan, O.T., Adetunji, C.O., Adeniyi, M.J., and Hefft, D.I. (2022). Computational Intelligence in IoT Healthcare. In: *Deep Learning, Machine Learning and IoT in Biomedical and Health Informatics.* CRC Press, p. 13. eBook ISBN 9780367548445. DOI: 10.1201/9780367548445-19

Oluwaseun, A.C., Phazang, P., and Sarin, N.B. (2017). Significance of Rhamnolipids as a Biological Control Agent in the Management of Crops/Plant Pathogens. *Current Trends in Biomedical Engineering & Biosciences*, 10(3): 54–55. Juniper Publishers Inc.

Oyedara, O.O., Adeyemi, F.M., Adetunji, C.O., and Elufisan, T.O. (2022). Repositioning Antiviral Drugs as a Rapid and Cost-Effective Approach to Discover Treatment against SARS-CoV-2 Infection. In: *Medical Biotechnology, Biopharmaceutics, Forensic Science and Bioinformatics*. CRC Press, p. 12. eBook ISBN 9781003178903. DOI: 10.1201/9781003178903-10.

Park, S.-J., Palvanov, A., Lee, C.-H., Jeong, N., Cho, Y.-I., and Lee, H.-J. (2019). The Development of Food Image Detection and Recognition Model of Korean Food for Mobile Dietary Management. *Nutrition Research and Practice*, 13(6): 521–528.

Patel, K. K., Khan, M. A., Kar A., Kumar, Y., Bal, L. M. and Sharma, D. K. (2015). Image Processing Tools and Techniques Used in Computer Vision for Quality Assessment of Food Products: A Review. *International Journal of Food Quality and Safety*, Vol. 1. 1–16.

Predrag, P., Lorenzo, J.M., Barba, F.J., Roohinejad, S., Jambrak, A.R., Granato, D., Montesano, D., and Kovacevi, D.B. (2018). Novel Food Processing and Extraction Technologies of High-Added Value Compounds from Plant Materials. *Foods*, 7(106): 1–16. doi:10.3390/foods7070106

Pushpavalli, M. (2019). Image Processing Technique for Fruit Grading. *International Journal of Engineering and Advanced Technology (IJEAT)*, 8(6): 3894–3897. DOI: 10.35940/ijeat.F8725.088619.

Qiu, C.(2016). Use of Extraction Technologies in Food Safety Studies. *Theses and Dissertations*, 987. http://openprairie.sdstate.edu/etd/987.

Ran, Anran, and Cheung, C.Y. (2021). Deep Learning-Based Optical Coherence Tomography and Optical Coherence Tomography Angiography Image Analysis: An Updated Summary. *Asia-Pacific Journal of Ophthalmology (Phila)*, 10: 253–260.

Rohit, S., Kumar, M., and Alam, M.S. (2021). Image Processing Techniques to Estimate Weight and Morphological Parameters for Selected Wheat Refractions. *Scientific Reports*, 11: 20953. https://doi.org/10.1038/s41598-021-00081-4

Sangeetha, J., Hospet, R., Thangadurai, D., Adetunji, C.O., Islam, S., Pujari, N., and Al-Tawaha, A.R.M.S. (2021). Nanopesticides, Nanoherbicides, and Nanofertilizers: The Greener Aspects of Agrochemical Synthesis Using Nanotools and Nanoprocesses Toward Sustainable Agriculture. In: Kharissova, O.V., Torres-Martínez, L.M., Kharisov, B.I. (eds), *Handbook of Nanomaterials and Nanocomposites for Energy and Environmental Applications*. Springer, Cham. https://doi.org/10.1007/978-3-030-36268-3_4

Shiv, R.D., and Jalal, A.S. (2015). Application of Image Processing in Fruit and Vegetable Analysis: A Review. *Journal of Intelligent Systems*, 24(4): 405–424. DOI 10.1515/jisys-2014-0079

Suganya, S. (2015). Analysis of Feature Extraction of Optical Character Detection in Image Processing. *International Journal of Engineering Research & Technology (IJERT)*, 3(4): 1–8.

Ukhurebor, K. E., Adetunji, C.O., Bobadoye, A.O., Aigbe, U.O., Onyancha, R.B., Siloko, I.U., Emegha, J.O., Okocha, G.O., and Abiodun, I.C. (2021). Bionanomaterials for Biosensor Technology. *Bionanomaterials. Fundamentals and Biomedical Applications*, 5–22.

Ukhurebor, K.E., Mishra, P., Mishra, R.R., and Adetunji, C.O. (2020). Nexus Between Climate Change and Food Innovation Technology: Recent Advances. In: Mishra, P., Mishra, R.R., and Adetunji, C.O. (eds), *Innovations in Food Technology*. Springer, Singapore. https://doi.org/10.1007/978-981-15-6121-4_20

Ukhurebor, K.E., Nwankwo, W., Adetunji, C.O., and Makinde, A.S. (2021a). Artificial Intelligence and Internet of Things in Instrumentation and Control in Waste Biodegradation Plants: Recent Developments. In: Adetunji, C.O., Panpatte, D.G., and Jhala, Y.K. (eds), *Microbial Rejuvenation of Polluted Environment. Microorganisms for Sustainability*, vol 27. Springer, Singapore. https://doi.org/10.1007/978-981-15-7459-7_12

Valous, N.A. and Sun D. W. (2012). *Image Processing Techniques for Computer Vision in the Food and Beverage Industries*. Woodhead Publishing, Chapter 4, 97–129.

Velesaca, H.O., Suarez, P.L., Mira, R., and Sappa, A.D. (2021). Computer Vision Based Food Grain Classification: A Comprehensive Survey. *Computers and Electronics in Agriculture*, 187, 106287.

Yi, H., Chen, Y., Qin, Z., Zhang, J., and Zheng, Z. (2019a). Optical Convolution Based Computational Method for Low-Power Image Processing, *Proceedings SPIE 11136, Optics and Photonics for Information Processing XIII*, 111360N (6 September). doi: 10.1117/12.2527733

Yi, S.N., Xue, W., Wang, W., and Qi, P. (2019b). Convolutional Neural Networks for Food Image Recognition: An Experimental Study. In *5th International Workshop on Multimedia Assisted Dietary Management (MADiMa'19)*, October 21, Nice, France. ACM, New York, pp. 33–41. https://doi.org/10.1145/3347448.3357168

4 Artificial Intelligence and Automation for Precision Pest Management

Charles Oluwaseun Adetunji, Olugbemi T. Olaniyan, Osikemekha Anthony Anani, Abel Inobeme, Osarenkhoe O. Osemwegie, Daniel Hefft, and Olalekan Akinbo

CONTENTS

4.1 INTRODUCTION

The recent development in machine learning, computer vision, artificial intelligence and mechatronics have led to tremendous advancement in the growth and utilization of remote sensing technologies for detection of diseases and pests, weeds, and making plant disease-management decisions. They give a perfect channel for rapid development of intelligent agriculture systems, most especially those having diverse precision utilization. Machine learning and artificial intelligence have been identified as two major promising areas in the fields of robotics, computer science, and automation. Machine learning involves the utilization of artificial intelligence that is built on the concept of a machine, whether as a microcontroller or computer, that can acquire knowledge from data and detection patter programs. The techniques in which a computer could detect from data without prior programing and switch to novel input so as to get some specific task done is known as machine learning. This process has been identified to require several amounts of data with regard to artificial intelligence. It has been discovered that both artificial intelligence and machine learning have several applications in modern-day agriculture (Ampatzidis, Bellis, and Luvisi 2017, Adetunji et al., 2020; Adetunji et al., 2020a,b,c; Oyedara et al.,

DOI: 10.1201/9781003207955-4

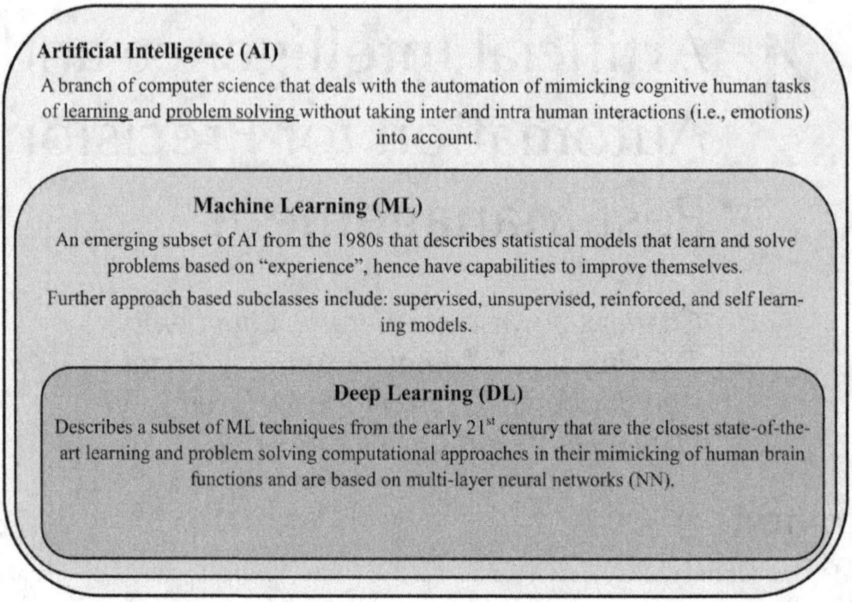

Artificial Intelligence (AI)

A branch of computer science that deals with the automation of mimicking cognitive human tasks of learning and problem solving without taking inter and intra human interactions (i.e., emotions) into account.

Machine Learning (ML)

An emerging subset of AI from the 1980s that describes statistical models that learn and solve problems based on "experience", hence have capabilities to improve themselves.

Further approach based subclasses include: supervised, unsupervised, reinforced, and self learning models.

Deep Learning (DL)

Describes a subset of ML techniques from the early 21st century that are the closest state-of-the-art learning and problem solving computational approaches in their mimicking of human brain functions and are based on multi-layer neural networks (NN).

FIGURE 4.1 Difference between Artificial Intelligence, Machine Learning and Deep Learning.

2022; Adetunji et al., 2022a,cb;d,e,f,g,h,i; Olaniyan et al., 2022). Figure 4.1 shows the difference between artificial intelligence, machine learning, and deep learning. Moreover, several transformations are presently documented in the agriculture sector, most especially through the invention of novel technologies that could enhance farm profitability and production yield (Himesh, 2018). The utilization of precision agriculture has played a tremendous role in the development of the modern agriculture revolution, which entails the first revolution with mechanization, the second revolution with genetic modification and, presently, with an increase in the application of farm knowledge, which might be linked to the emergence of available agricultural data (Zhang, 2019). This has been shown to increase operating profits and net returns (Schimmelpfennig, 2016; Panda, 2020).

It has also been observed that the introduction of these novel technologies have enhanced the preservation of an ecofriendly environment and sustained food security. Moreover, there is a need to introduce to farmers training in the application of relevant artificial intelligence (AI), disseminate information via an efficient extension service network, and readily fund farmers so as to guarantee adequate adoption of sustainable AI-based farming technologies. The merits of these technologies demonstrated in enhanced production, quantity, and quality of food, reduction in the prices of farm produce and sustenance of ecofriendly practices (Díez, 2017).

Pests and disease have been identified among the major biotic factors that mitigate against the increase in the mass production and profit from agricultural produce. The application of synthetic pesticides has led to rapid emergence of resistance in agricultural pests and pathogens, higher intoxicantion or contamination of crops by pesticides residue, depletion of soil nutrients and of beneficial microbiome. Therefore, there is a need to deliberately search for sustainable techniques that could boost the level of agricultural produce, most especially, and help feed the ever-increasing global population (Adetunji et al., 2017a,b; Adetunji et al., 2018, Adetunji et al., 2019).

Therefore, this review provides detailed information on the role of artificial intelligence and automation in precision pest management.

4.2 PROCESSES INVOLVED IN THE IDENTIFICATION OF PESTS AND DISEASES AND IMPLEMENTATION OF THE INTEGRATED PEST MANAGEMENT

Sreelakshmi and Padmanayana (2015) reported that early detection of pests and pathogens has been a major issue in the agricultural sector, with excessive utilization of pesticides causing serious environmental and health concerns. Thus, an integrated pest-management system that combines different approaches may be the way out of this difficulty. Recently, studies revealed that application of machines and digital image techniques in the agricultural sector have drastically improved food production due to increased capacity for plant protection and crop management (Singh et al., 2019; Debauche et al., 2020, Panda, 2020). These techniques involve the use of infected leaves, stem, root and fruit images uploaded into a digital camera, processed in a system computer, extracted, and analyzed through image classification techniques (MATLAB software and Support Vector Machine classifier). This approach helps to reduce the use of pesticides and promotes quick preventive measures, thereby improving crop management systems. Yiannis et al. (2017) reported that increased advancement in AI technology and rapid development of automated technologies have transformed the agricultural sector. The adoption of cloud-based solutions, Wireless Sensor Network, Internet of Things (IoT), and the Internet provides a huge opportunity for improving forestry, agriculture, and urbanized farming (Dbauche et al, 2020). Some of these technologies – machine vision, laser technologies, machine learning, global positioning systems, artificial intelligence, mechatronics and actuators – are currently adopted for precision agriculture, irrigation, harvesting, pest- and pathogen-control systems, weeding, and crop management. Plant disease symptoms and images have been successfully developed and interpreted through machine diagnosis into data acquisition for crop-disease and pest-management. A robotic crop-pest diagnostic system uses biosensors and image analysis to rapidly detect and monitor early pathogens, disease symptoms, abiotic stress and weeds in seedlings, leaves, fruits and post-harvest crops. Yiannis et al. (2017) observed that the robotic eye is by far more sophisticated and more efficient than the human eye or laboratory procedures, and could serve as an experienced pathologist.

Banana production, a popular marketable fruit crop in most parts of the developing countries, is constantly faced with pests and diseases, thus negatively affecting productivity. The timely development, detection and early monitoring of the onset of crop disease through artificial intelligence could support the banana farmer's productivity (Michael et al., 2019). Thus, different datasets now exist of images on pre-screened banana pest and disease symptoms in India and Africa. These datasets were used to develop detection models through various convolutional neural networks that lead to the development of eighteen different models. Furthermore, the models generated 90 percent accuracy in all the tested experiments as compared with others found in the literature. Automated disease-detection methods' validity and performance also produced robust digital banana disease- and pest-detection accuracy with a significant success rate. It is therefore logical to suggest that many models be developed for different crops, such as bean, brachiaria, potato and cassava, utilizing convolution neural networks and drones for multiple disease detection.

Similarly, robotic technologies' deployment to the agricultural sector, which is the fastest growing industrial sector charged with producing high-quality and yield crops, has helped achieved significant early detection and monitoring of pests and pathogens to minimize economic loss. The robotic technology in agriculture consists of about five components that include controller, sensors, actuators, arms, and effectors (Hassan et al., 2016). The robotic technologies ensure progressive sustainable agriculture and food security through pest detection, pathogen management control and identification, weeding, micro spraying, weed mapping, seeding, insect pest traps with low-power image sensors, harvesting, and irrigation. Durgabai et al. (2018) reported that climate and disease status play a significant role in the productivity and yield of many agricultural crops across the globe. A majority of the global population depends on agriculture, thus cultivation of important crops requires serious attention for optimum quality and yield. This outcome can be achieved through the deployment and utilization of artificial intelligence for early detection and monitoring pests, diseases and different climatic conditions. Crop science and practice involve the utilization of various techniques that include machine learning, algorithms, and the Internet of Things to improve crop production, and pathogens and pest management. Barbedo (2019) observed that many pest outbreaks are very unpredictable; thus early detection, monitoring, and effective management become possible through the use of modern agricultural technologies. Advanced imaging technologies are used as noninvasive crop-monitoring methods for the early detection of pests and pathogens in agricultural systems. Again, in precision agriculture, drones are now becoming a popular technology being deployed to promote crop health through monitoring, early detection, and response to pests and pathogens.

In Kenya, agriculture contributes a significant percentage (24.5) to the national gross domestic product, making this sector one of the key economic hubs, with maize occupying the central position. Many farm crops suffer attack from diseases and pathogens during the planting season, thus reducing the farmers' economic gains. In the face of the numerous challenges facing this sector, experts have alluded to the fact that delivering technological approaches like artificial intelligent systems, algorithms for early detection, diagnostics, and monitoring will increase the output and productivity of farmers in Kenya (M'mboyi et al., 2010; Ngugi et al., 2020). Across the globe, reports have affirmed the propensity of many of these pests and pathogens to cause extreme loss and nutritional deficiency while posing a challenge to many farmers. Consequently, image processing techniques and deep learning have been recommended for the early detection and monitoring of diseases to deliver ultimate performance in agro-based industries.

Fernandez-Quintanilla et al. (2018) gave an account of deep-learning-based target detection and the smart-sprayer approach in precision agriculture. This is currently being utilized for pest and weed control, while also rapidly driving transformation in the agricultural sector globally. In addition, the deployment of technologies for automation farming has also shown potential in the improvement of food yield and quality (Nikhil et al., 2020). Food, one of the basic needs of humanity, is being constantly threatened by different factors like pests and pathogenic attack, extreme weather conditions, drought, and anthropogenic activity. Early pest detection and surveillance of crop pathogens utilizing artificial intelligence for diagnostic of different crop symptoms constitute an emerging technology that will change the face of the agricultural sector by increasing the production of food. Akshata et al. (2019) observed that in order to meet the consumer demands on food production due to the increased population explosion, there must be increased investment and upgrading of current infrastructural deficits in the agricultural sector. This can be achieved by deployment and utilization of artificial intelligence, artificial neural networks, machine learning, satellite imagery, Bayesian belief network, Cloud Machine Learning and Information Regression Analysis to monitor unpredictable soil, crop health, and pathogenic control measures in smart farming.

According to Diego and Rafael (2018), the global economy can be enhanced through huge investments in grain production. This can be done via serious deployment of artificial intelligence algorithms and tools for precision agriculture. Grain productivity – maize, wheat, rice, barley, soybean – and other products such as orange, almond, papaya, lemon, potato, and corn has been threatened across the globe in the past five years due to unprecedented pest and pathogenic attacks. These attacks have drastically reduced yield and quality, resulting in food shortage and resulting insecurity, a situation correctable by advanced artificial intelligence techniques such as Deep Belief Networks. This is an important computer vision method suggested to produce accurate descriptive datasets. Owomugisha and Mwebaze (2016) also carried out crop-disease diagnosis through automated farming, which is an important task utilizing leaf images for dataset generation and analysis. The most important part of this automated approach based on artificial intelligence is the offered degree of sensitivity or early detection before the plant becomes symptomatic and affords a strategy of using the early opportunity to rapidly apply treatment. Therefore, research and innovation in agriculture have ensured the development and deployment of technologies amenable to many challenges ravaging the sector. Robots are now being utilized for crop and animal sensing of pests and pathogens, weather conditions, drought, harvesting, planting, and weeding. Increasing global artificial intelligence penetration along with recent technological advancement in computer vision have made it possible to utilize deep learning to improve crop-disease diagnosis (Balamurugan et al., 2016).

An artificial neural network system for pest insect identification and monitoring can improve food production and economic gains via the use of digital image analysis to generate datasets (Peter et al., 2009). Vreysen et al. (2007) noted that an integrated pest-control system is essential for transforming the agricultural sector and remediating several diseases that hamper development and productivity in livestock farming

(Suresh et al., 2019). Many of these diseases can be detected through the adoption of technologies such as artificial intelligence. Verónica and Francisco (2020) noted that advanced data management in smart farming helps farmers make quick decisions on how to maximize productivity and sustainability. Therefore, computer vision, artificial intelligence, machine learning, deep learning and mechatronics are receiving more attention as tools in recent technological advancement for development of intelligent agricultural systems (Abdulridha et al., 2018). These systems utilize datasets to analyze and identify plant diseases and pests, plus symptoms, particularly in *Olea europaea* L (olive tree), citrus, and grape Mack et al. (1998). Adejumo (2005) and Puig et al. (2015) affirmed that crop pests annually cause millions of dollars of losses in crop yield and productivity across different developing countries, supporting the need for the early monitoring and detection of pests in order to reduce losses and costs and stem quality devaluation. The utilization of remote sensor technology can rapidly impel biosecurity and precision agriculture in the sorghum crop. A UAV platform can detect beetles feeding on the roots of plants, generate high-resolution RGB imagery for further analysis through different algorithms to deliver automated early pest crop damage assessments, monitoring, and biosecurity surveillance.

Globally, the utilization of pesticides has grown tremendously in the last 20 years from 3.5 billion to 45 billion kg/year, accounting for many environmental and human health hazards (Jules and Zareen, 2015). While this has encouraged the adoption of integrated pest management systems – to protect the environment, human health, food quality, and crop yield by reducing dependency on synthetic chemicals – pests and other crop pathogens still posed a huge danger to agricultural livelihoods, food security, and the attainment of sustainable development goals of low poverty or strategies for poverty alleviation . Consequently, farmers in Asia and Africa have reduced dependency or over-reliance on pesticides through the adoption of integrated pest management control strategies. Integrated pest management is being promoted in the United States to advance sustainable agriculture, and it has helped many farmers in saving capital, improving the environment, human health, and crop productivity (Molly et al., 2011). Although integrated pest management is affirmed as an effective plant disease-management approach, it has been slow to spread to mainstream agriculture due to skepticism associated with its strategic cost implication and effective field application.

4.3 UTILIZATION OF SMART DIGITAL DEVICES IN THE PREDICTION OF THE PLANT HEALTH STATUS AND THE ENVIRONMENT

Digital agriculture has become the main farming technique employed in modern farming intensification. Shamshiri et al. (2018) remarked that a novel research-and-development (R and D) approach to farming robotics is effective, affordable, and complex. Suffice it to say that digital agriculture requires multidisciplinary teamwork or contributions from different sectors such as crop management, system integration, software design, instrumentation and sensors, deeply intelligent learning systems, dynamic control, mechatronics, computer science, engineering, and horticulture. Furthermore, digital agriculture is a novel, recent, and versatile technique that depends on data analysis, robotics, and sensors for effective functional delivery. This method can be employed in the control of harvesting, scouting the field for early signs of pests, pathogens, and weed infiltrations using high sensor optimization, digitalization, algorithms, and object documentation tools. In agriculture, robotics is used in harvesting and weeding, and can perform tedious work more efficiently and faster than humans and other traditional methods based on automation time setting. The SWEEPER robotics is a form of technology that has been employed in the picking of economically valued fruits such as apples and citrus harvested massively for the juice trade. This technique aids in the collection, detachment, attachment, and locomotion of fruits as well as in the reshaping and pruning of leaves and plants, thus enhancing efficacy. The modification of an extant mechanized harvesting system for collective robotic functionalities can serve as a more promising tool when compared to a solitary robotic system. It has, however, been hypothesized that the planting and breeding of crops using both humans and robots remains a fundamental form collaboration in resolving problems relating to food security and safety in greenhouses and fields' farming to resolve problems related to the health and condition of the system.

Weiss and Biber (2011) evaluated a 3D FX6 LIDAR robotic sensor in the mapping and detection aspects of the management of farming activities. The 3D FX6 LIDAR is a traditional sensor equipped with properties such as stereo vision and mobility . It is mostly used in agricultural activities requiring the segmentation and detection of plant qualities and ground profiling via automation navigation, mapping, and localization, even in dusty and foggy environments. Some real-time algorithms have been programmed to detect and single out plants in vague climatic conditions. The accuracy of the robot even in unfriendly weather condition in the localization and detection of unhealthy plant situations makes it a more reliable, high-resolution sensor. The detection of the individual plant using this approach also offers future promise for weeding, fertilizing, and spraying fields with pesticides at a cheaper cost, thereby leading to a dramatic improvement in economic efficiency operations and bio-safety of lives. Additional studies are therefore suggested on the identification of crops by rows, using similar technology, and the future promise for agriculture. Moreover, machine learning (ML) techniques should be made amenable to discriminate between different types of plants and to learn novel strains autonomously when used in farm operations. These techniques could also be used to enhance the algorithm clustering through the learning principles of cluster split-offs.

Michelmore et al. (2017) evaluated the computational technology involved in the translation, foundation, and enhancement of the health of plants while confirming plant health as one of the fundamental parts in maintaining their productivity, quality, and economic values. Plant health also has very strong implications for meeting human food needs and giving impetus to crops' fast growth. Worldwide, food security is being threatened by climate change, environmental stress, weeds, and pathogens. However, for the first-time novel computational and analytical technologies have provided a solution at different stages of population growth. Its direct or indirect detection have influence at the organismal, cellular, and molecular levels of weeds, beneficial microbes, pests, and pathogens of plants requires further investigation. Monitoring using standard crop-health techniques like stem rust should be employed to reduce the vulnerability of crop supply to biotic components. The strategic automatic interventions, in conjunction with chemical and biological protectants, are now recommended tools in mitigating weeds, pests, and pathogens that have posed limitations to many agricultural communities. Genetic editing and green technology are novel methods that are becoming attractive for use in evaluating the health of crop plants in many developed nations.

Machine learning (ML) techniques used in the discovery of resistant genetic components linked with brown rust disease in sugarcane. Sugarcane has a complex genomic intricacy, which has delayed its development into genetic and molecular breeds of different cultivars. However, assorted marker-assisted resistant genes for brown rust in sugarcane plant have been discovered successfully, but need to be further investigated using advance engineering and ML approaches. In this study, the authors utilized genotyping sequencing by full-sib offspring relating its genetic components with the phenotypes of the brown rust disease of the sugarcane plant. The use of ML using SNPs multiplex polyploid data to predict the status of phenotypes showed 14,540 SNPs, which attained about 50 percent accuracy average prediction employing various models. The outcome of the algorithm from the dataset which was utilized to assess the accuracy of the prediction revealed about 95 percent accuracy with 131 SNPs of the dataset linked to the sugarcane disease at the quantitative trait loci (QTL), auxiliary and regional genes. This suggests that great potential can be derived from this method of recapturing resistant genes in related plants in the war against various diseases of crops. The approach can also be used to predict resistant genes in plants that can potentiate the health of plants via the use of multiplex polyploid data (Aono et al., 2020).

Cruzan et al. (2016) studied the application of small drones; micro-UAVs in plant ecology and their use in the survey of plant community in low-elevation regions. While the small aerial robotic drone can also be used in the mapping of small-to-medium vegetation, this study modified it into a DSM (digital surface model) with clear orthomosaic images combined with automated and manual composites to map vegetation within a specified range. The accurate confirmation of the vegetation habitat and improved background of the estimated species coverage range were forecast, supporting further the use of mechanized robotic drones in the health and welfare study of vegetation with future promise for use in other highly vegetation-covered regions of the world.

Conversely, Araus et al. (2018) emphasized the benefit that could be derived from the utilization of genetic breeding in plant programming via the translation of enhanced phenotyping models and inefficiency of the implementation of the enhanced phenotyping models. The effective field application of this approach in the monitoring of plants required the integration of, phenotyping with a larger context in data management, assessment platforms, instrumentations, and selection of desirable plant traits. This implies that phenotyping is versatile in content and usage. Beyond that, it can also be employed as a tool for crop modeling, population prediction, experimental management, and a 3-D variability tool. Phenotyping models are becoming attractive as a tool that can be used to bridge the space between phenotypers and breeders when implemented efficiently in the field.

Rudic et al. (2020) compared and evaluated the drone and smartphone Lidar techniques used in the characterization of spatial differences in the plant area index (PAI) of a humid forest. The assessment and estimation of the LAI (leaf area index) across the varied landscape in the tropical rain forest is important in several ecological investigations. Numerous indirect and direct methods used in the estimation of LAI can be compared and developed. Nonetheless, these techniques are time-consuming, require expertise and can be exorbitantly expensive. Furthermore, utilizing CAN-EYE; a free IPS (image processing software) as an effective PAI confirmed that a smartphone effective PAI of 0.1-4.4 at $r = 0.62$, $n = 42$, $p < 0.001$ significantly relates to the effective variation and the smartphone. This association was further enhanced when the secondary and old forests' growth were assumed to possess a varied angle distribution of leaves for the pianophiles and spherical algorithms correspondingly at $r = 0.77$, $n = 42$, $p < 0.001$. However, some eccentricities in the effective PAI magnitude estimations, which depended on the IAO (image analytical options) were noticed. This suggested that the images from the smartphone can be utilized to characterize the 3-D variation in the effective PAI in a multiplex, varied forest upper story or canopy, with only minor decreases in the descriptive power as related to the real digital curved photography.

The use of aerial images in predicting the kernel dimensions, yield, and flowering time of *Zeamays* was also valuable in analyzing images for gauging the phenotypes of the crop (Wa et al., 2010). This may substitute the native measurements, since it is more reliable and an efficient capture of similar information. The spectrally modified consumer-grade photographic camera was used as the aerial vehicle in the captured images and depends on the value of the pixel in the form of a normalized BNDVI (blue and near-infrared wavelength bands). BNDVI, however, improves in genotype imageries early in the plant season. This accuracy later declines with altered image shape as the season climaxes, as plant canopies cover the soil, which becomes extensively green and senesced. The PCA (Principal Components Analysis) carried on the dataset, showed a change in the shape of the histogram at PC1, PC2, and PC3 (canopy closure, BNDVI dispersion, and spread of the distribution), were observed respectively. The flowering timing correlation was observed in PC3 and PC2 at $r \approx 0.5$ a few days before the time when 50 percent of the *Zeamays* showed tassels. To quantify the dimensions of the kernel, three ears were selected using image analysis. They were harvested mechanically to determine the weight of the grain and the yield of the plot. It was observed that the correlation of the measurement of PC2 and the yield dropped in June but later improved at 0.4 on day 10 after the maize plant flowered. However, the length of the kernel correlated in the same line with PC2. The relationship between the thickness of the kernel and the PC2 also showed a similar but upturned period course. These outcomes revealed that more mid-season values of the BNDVI correlate + with the thin and tall kernels. The results of the regression analysis carried on the BNDVI predicted the rate of yield and flowering as $r = 0.4–0.69$ and $r = 0.54–0.79$ respectively. The authors stated that both software and hardware were readily available to demonstrate the effectiveness of the phenotyping modeling used in the expression and prediction of the dimension of the kernels, the yield, and the flowering rate of the maize plant. The incorporation of these methods can serve as a future tool for imaging the breeding and genetic programs of maize and related grains to enable sustainable health and effective monitoring of the community of grains.

Another novel spectral plant indices in the distant detection of phytomass in grassland vegetation communities is, according to Vescovo et al. (2012), the NIR (near-infrared) berm wavelengths mechanism, which measures ISI, NIDI, and NDSI (infrared slope index, normalized infrared difference index, and normalized difference structural index) with RV (reflectance values) bands of 863–881 nm (H25), 745–751 nm (H18), and mode 5 (Chris Proba). The assessment of the phytomass and reflectance hyperspectral of vegetation sites in the Austrian and Italian mountains, utilizing a hand-held spectro-radiometer tool, showed a full-plant story cover, strong overload, and many native vegetation indices, NDVI, EVI, EVI 2, RDVI, and WDRVI (normalized difference vegetation index, modified simple ratio, enhanced vegetation index, enhanced vegetation index 2, and renormalized difference vegetation index). On the other hand, NDSI and ISI were correlated linearly to the vegetation phytomass with insignificant inter-yearly variability. The relationships between the phytomass, NDSI, and ISI were nevertheless site-specific, which was based on the WinSail model and might be due to the vegetation reflectance and plant species structure. Further studies may therefore be necessary to verify the importance of utilizing the multispectral sensors or indices for monitoring grassland composition and the biological and physical variables on similar ecosystems, as well as testing these associations with satellite and aircraft sensors information. NDSI and ISI could therefore serve as promising non-destructively temporal monitoring tools in phytomass vegetational variability.

Meshesha et al. (2020) predicted the health status of grassland biomass in Somali zones using satellite imagery. In the Somali zone, livestock farming is a centuries-long system of farming that has become part of the livelihoods of people around that region. This region is normally drought susceptible. The authors downloaded 2 sentinel images and manufactured different spectral data of near-infrared, red, and blue bands, and estimated the NDVI and EVI indices, which were used to develop 55 plots or sampling forecasting zones. The results from their study showed improved plant biomass was correlated significantly with both NDVI and EVI at $R2 = 0.81$; $P < 0.001$ and $R2 = 0.87$; $P < 0.001$, correspondingly. Both indices gave a favorable vegetational biomass prediction in the studied zones. The models also indicated local estimates of the EVI and NDVI at 0.78 t/ha of 39 792 t/year and 0.76 t/ha of 38 772 t/year respectively. An estimated spatial variability of 0.22–4.89 t/ha. year in the vegetational biomass was observed. The authors suggested a vegetation biomass satellite description for grasses within two months after the wet season or before the estimated and evaluated they get to the harvest time or matured.

The performance of satellite imagery in GPP (gross primary production) using the FLUXNET model across the world was studied by Huang et al. (2019). The satellite-sourced VIs (vegetation indices) have been globally utilized to estimate the GPP, even though its relationship fluctuates with the BRDF (bidirectional reflectance distribution function), timescales, biomes, and the extant indices used. While it was noticed that Vis could include NDVI, EVI, and EVI 2 alongside the MODIS (moderate resolution imaging spectroradiometer), the vegetation indices elucidated lesser variance in the tower gross primary productivity at the yearly scale compared to the regular (once a month) scale. Whereas the vegetation indices were affected by water stress and temperature, the Vis is less susceptible to water stress than temperature. It was also noticed that environmental factors in combination with the Vis enhanced the prediction of the gross primary productivity than the vegetation indices alone. Therefore, the association of the VI-GPP indices can help improve the influence of the relationship of BRDF, timescales, and biome in further studies.

Pettorelli et al. (2005) utilized NDVI-based satellites to evaluate the responses of ecological changes in the environment. Changes in the environment affect the dynamics and distribution of animals and vegetation populations which is significant for plant ecologists to better predict the influence of habitat degradation, reduction in biodiversity, and global warming. The capacity to predict the responses of ecological factors in the environment has often been obstructed by the knowledge of the interactions of different trophic levels. However, NDVI has drastically changed this in both spatial and temporal scales. The utilization of the NDVI has aided in the enhancement of ecological studies, which has the potential to unlock the role of each trophic interaction of organisms and the community function(s) relative to abiotic and ecosystem structure context.

With the advancement of IT (information technology), UAVs (Unmanned Aerial Vehicles), IOT have been utilized more as the monitoring tools in the field to investigate the incidences of pests and diseases on a hectare of agricultural land (Gao et al., 2020). IOT-based technology can source for real-time data parameters of the crop development and growth via numerous low-cost sensors equipped with high-definition smart digital cameras to capture images that show the occurrences of the invasion of diseases and pests of the crops. The IOT devices can track long-distance problems on the farmland. However, based on inadequate energy additives, a tracking device using solar energy is used to complement the energy and harvesting timing of the automated device. The issues of the low flight rate of the UAV are corrected by a flight mode which ensures the optimum use of extended flight rate and wind force. The issues of image capturing are addressed using cloud data for evaluating the rate of damage of diseases and pests due to the spectral analysis technology. The authors reported that susceptibility to pests and diseases can only occur when the temperature is at the range of 14–16 °C, and with an increase, rainfall reduces the proliferation of the disease wheat powdery mildew.

4.4 INNOVATIVE SENSORS, SATELLITES, DRONES AND ROBOTS THAT CAN HELP IN THE CAPTURING OF THE PICTURE OF PLANT HAVING DISEASES AND PESTS

Geographic information system, global positioning systems as well as different sensors are vital in the area of precision agriculture. They emphasize the indispensable role of robotics and satellite remote sensing in monitoring various activities of pests and plant diseases. Bansod et al. (2017) described the application of unmanned vehicles in remediating the challenges associated with ground-based conventional systems. In furtherance of innovating this technology, Sanchez et al. (2019) used computational vision science based on artificial neural network (ANN) for the quantification of effect of *Jacobiasca lybic*, a pest common in vineyards. This combination improved the segmentation of image in the affected plants' vegetation, affirming the higher efficiency of this method in the accuracy assessment of pests' impact through the removal of soil effects.

López and Mulero-Pázmány (2019) evaluated the potentials of using drones in monitoring protected or conserved areas. The authors state that there is a clarion call to protect conserved zone areas of the world due to the incidence of environmental degradation that have threatened the biodiversity therein. Drones have been used recently to monitor natural and human activities or influence protected as well as inaccessible zones to better predict and manage the negative outputs that might ensue from the zone. The potentials of drones in these zones could be more effective if combined with novel smart and digital technologies. This technology is valuable for key surveys involving five distinct areas such as environmental disaster and management, ecotourism, law implementation, monitoring of the ecosystems (particularly terrestrial), and wildlife monitoring where it helps in identifying specific challenges and gaps in the monitoring research. Drones therefore play a key role in the effective conservation and management of vegetation by addressing analytical and operational shortcomings in protected zones of the world.

Agroclimatic-based powdery mildew mitigation required the setting up of climate models derived from data gathered from simple climatic devices coupled with satellites (Ma et al., 2018). Everitt et al. (2003) also applied data from remote sensing gathered from satellites and manned aircraft in the management of pests. This gave promise to the use of videography and aerial photography in the detection of infestation by arthropods in farmlands.

Stanton et al. (2017) used a multispectral sensor attached to a fixed-wing drone for the counting of aphids throughout the growing season and verifying the infestations on the ground. Furthermore, the monitoring technology used in agriculture such as satellites, robotics, and drones amongst others, have the unique merit of good timeliness, brief period of revisiting, affordability and being amenable to a broad area (Li et al., 2020). Elijah et al. (2018) reviewed that the emergence of pests and diseases is closely linked to changes in climatic conditions within the time of cultivation and growth, hence, satellite-based technology is useful in the collection of basic data on the entities needed, thus providing an easy approach in the monitoring of the growth process in a real time pattern.

Dimitrio et al. (2016), in a related study, put forward a hardware design framework of actuators networks and wireless sensors for practical agricultural applications that would be relevant in curtailing agricultural pests and diseases. Ballesteros et al. (2014) documented that in the area of agriculture, a remarkable potential has been displayed by UAVs as aerial tools for the assessment and monitoring of crops, determination of height of agricultural crops, mapping of weeds, and detection of plant diseases amongst others.

In another study, Joalland et al. (2018) did an examination on the derivation of hyperspectral, spectral, and temperature data from a UAV-borne, and handy, sensor for the discrimination of sugar beets cultivars that are vulnerable or capable of withstanding nematodes. The authors were able to determine the most useful traits for the task based on validity. Filho et al. (2020) in their work reported two kinds of precision pest management, which include reflectance based components, such as airborne, ground-based, or remote sensing using orbital, and the second, which is the precision control system that involves various distributors of pesticide spray rigs and natural enemies. They explained that both technologies could be implemented using manned or unmanned components.

Filho (2019), in their study, used a drone-based sensor, multispectral in nature, as well as an hyperspectral sensor for the assessment of the impact of stress brought about by arthropods in soybean farms together with other pests, such as caterpillars, stink bugs, and whitefly. Barbedo (2019) outlined a detail of remote sensing investigations that are drone-based for various uses such as monitoring and detection of diseases causing microorganisms, pests, and droughs as well as deficiencies of mineral nutrients. He emphasized that there is an increasing use of drones in remote studies, which are highly efficient cost-wise in the inspection of smaller fields.

Faical et al. (2014) was of the view that drone usage in agricultural pest and disease management has several merits, which include tendency of reduction in spray drift, minimal exposure of users to the chemicals, and an easier deployment strategy. Judd and Gardiner (2005) documented that one of the novel applications of drones in the area of management of pests and diseases is the discharge of sterile insects. Codling moth, which is an example, is a major challenge in apple orchards (*Mulus domestica*).

Shamshiri et al. (2018) emphasized the role of simple scale drones and robots that would collaborate in the optimization of agricultural inputs relevant to the management of agricultural pests. They emphasized that while the drones have become reliable components of modern agriculture, it will not however be realistic to expect a completely automated agriculture in the future. Shafi et al. (2019), in their review, presented remote and near sensors network and their relevance in modern-day farming. They explored wireless nodes, sensors and communication technology in their role to obtain spectra photographs of agricultural crops as well as the vegetation indices for analyzing the spectra images. They proposed Information of Thing (IoT) based smart solutions in the monitoring of plant health and diseases.

Mattupalli et al. (2017) reported their work on the use of unmanned aircraft systems (UAS) in investigating the spread of phymatoctrchopsis root rot (PRR) disease, revealed through the use of high-resolution aerial imagery. They considered a 50.4 hectare with a center pivot irrigation system. Aerial imageries were captured by the UAS as it navigated along a global positioning system (GPS) guided flight route. The researchers concluded that the number of images obtained was affected by factors such as height, spatial resolution desired, and the area to be photographed.

It has been established that there are several researches that could be channeled towards the early detection of diseases and control of pests using advanced technology such as satellites, drones and robotics. Moreover, various researchers have used sensors, fluorescence sensors, thermal sensors, and spectra sensors in the collection of imagery data. The spatial images were affected by the distance from the sensor to the object. Golhani et al. (2018) documented the use of hyper spectra images for the analysis of pest infestation and plant health through the use of unmanned and manned vehicles. The analyses of the images captured are done through the use of machine learning approaches for the identification of the disease. They opined that neural networks are useful for the processing of the imagery data as a result of their capacity for learning patterns and structures.

Rumpf et al. (2010) used four supervised algorithms for the spectra images for the early and quick detection of disease in the sugar beet plant. The indexes for multiple vegetation were calculated for the purpose of prediction. Sanghvi et al. (2015) carried out a study in which a technique of data mining was used on data already collected from rice paddy and wheat field in India. They used Sammon's mapping for the reduction of dimension and scaling from varying dimensions.

4.5 CONCLUSION

This chapter provides detailed information on the application of artificial intelligence and automation for precision management of diseases and pests. Specific accounts involving the use of artificial intelligence and automation for the identification of pest and diseases, assessment of their economic damage, implementation of integrated management, and the procedure involved in the assessment of the success of the integrated pest management strategy were provided (Adetunji et al., 2021a,b,c; Ukhurebor et al., 2021; Sangeetha et al., 2021; Adetunji and Ukhurebor, 2021; Oluwaseun et al., 2017; Adetunji et al., 2012; Arowora et al., 2012; Dauda et al., 2022a,b; Ukhurebor et al., 2021; Okeke et al., 2021; Adetunji and Anani, 2021; Nwankwo, et al., 2021; Adejumo and Adetunji. 2018; Ukhurebor et al., 2021). Moreover, heterogeneous ground and aerial robots, together with innovative sensors, satellite, drones, or robots, are some of the technologies that offer more accurate prediction of the plants' health status, of the environment, severity of disease incidence or spread, economic repercussions, and gauging the impact of evolving remedies. The chapter also covers many of these technologies that are now being increasingly adapted for agriculture use through enhanced end-effectors and better-quality decision control algorithms, Smartphone apps, normalized Difference Vegetation Index Satellite, drones, or robots.

REFERENCES

Abdulridha, J., Ampatzidis, Y., Ehsani, R., and de Castro, A. (2018). Evaluating the performance of spectral features and multivariate analysis tools to detect laurel wilt disease and nutritional deficiency in avocado. *Comput. Electron. Agric.* 155: 203–2011.

Adejumo, T. O. (2005). Crop protection strategies for major diseases of cocoa, coffee and cashew in Nigeria. *Afr. J. Biotechnol.* 4 (2): 143–150.

Adejumo, I. O., and Adetunji, C. O. (2018). Production and evaluation of biodegraded feather meal using immobilised and crude enzyme from Bacillus subtilis on broiler chickens. *Braz. J. Biol. Sci.* 5(10): 405–416.

Adetunji, C. O., Adejumo, I. O., Afolabi, I. S., Adetunji, J. B., and Ajisejiri (2018). Prolonging the shelf-life of "Agege Sweet" Orange with chitosan-rhamnolipid coating, *Horticul. Environ. Biotechnol.* 59 (5) 687–697. http//:doi:10.1007/s13580-018-0083-2

Adetunji, C. O., Afolabi, I. S., and Adetunji, J. B. (2019). Effect of Rhamnolipid-*Aloe vera* gel edible coating on post-harvest control of rot and quality parameters of "Agege Sweet" Orange. Agriculture and Natural Resources. *Agr. Nat. Resour.* 53: 364–372.

Adetunji, C.O., and Anani, O.A. (2021). Bioaugmentation: A powerful biotechnological techniques for sustainable ecorestoration of soil and groundwater contaminants.

In: Panpatte, D.G., and Jhala, Y.K. (eds.), *Microbial Rejuvenation of Polluted Environment. Microorganisms for Sustainability*, vol 25. Springer, Singapore. https://doi.org/10.1007/978-981-15-7447-4_15

Adetunji, C.O., Anani, O.A., Olaniyan, O.T., Inobeme, A., Olisaka, F.N., Uwadiae, E.O., and Obayagbona, O.N. (2021a). Recent trends in organic farming. In: Soni, R., Suyal, D.C., Bhargava, P., and Goel, R. (eds.), *Microbiological Activity for Soil and Plant Health Management*. Springer, Singapore. https://doi.org/10.1007/978-981-16-2922-8_20

Adetunji, C.O., Egbuna, C., Oladosun, T.O., Akram, M., Michael, O., Olisaka, F.N., Ozolua, P., Adetunji, J.B., Enoyoze, G.E., and Olaniyan, O. (2021b). Efficacy of phytochemicals of medicinal plants for the treatment of human echinococcosis. Ch 8. In: *Neglected Tropical Diseases and Phytochemicals in Drug Discovery*. Wiley. DOI: 10.1002/9781119617143

Adetunji, C. O., Fawole, O. B., Afolayan, S. S., Olaleye, O. O., and Adetunji, J. B. (2012). An Oral Presentation during 3rd NISFT Western Chapter Half Year Conference/General Meeting, Ilorin, pp. 14–16.

Adetunji, C.O., Michael, O.S., Nwankwo, W., Ukhurebor, K.E., Anani, O.A., Oloke, J.K., Varma, A., Kadiri, O., Jain, A., and Adetunji, J.B. (2021c). Quinoa, the next biotech plant: Food security and environmental and health hot spots. In: Varma, A. (eds.), *Biology and Biotechnology of Quinoa*. Springer, Singapore. https://doi.org/10.1007/978-981-16-3832-9_19

Adetunji, C.O., Mitembo, W.P., Egbuna, C., and Narasimha Rao, G.M. (2020). Silico modeling as a tool to predict and characterize plant toxicity. In: Mtewa, A.G., Egbuna, C., and Narasimha Rao, G.M. (ed.), *Poisonous Plants and Phytochemicals in Drug Discovery*. Wiley Online Library. https://doi.org/10.1002/9781119650034.ch14.

Adetunji, C.O., Nwankwo, W., Olayinka, A.S., Olugbemi, O.T., Akram, M., Laila, W., Samuel, M.O., Oshinjo, A.M., Adetunji, J.B., Okotie, G.E., and (Diuto) Esiobu, N. (2022a). Computational intelligence techniques for combating COVID-19. In: *Medical Biotechnology, Biopharmaceutics, Forensic Science and Bioinformatics*. CRC Press, p. 12. eBook ISBN 9781003178903. DOI: 10.1201/9781003178903-16

Adetunji, C.O., Nwankwo, W., Olayinka, A.S., Olugbemi, O.T., Akram, M., Laila, U., Olugbenga, M.S., Oshinjo, A.M., Adetunji, J.B., Okotie, G.E., and (Diuto) Esiobu, N. (2022b). Machine Learning and Behaviour Modification for COVID-19. In: *Medical Biotechnology, Biopharmaceutics, Forensic Science and Bioinformatics*. CRC Press, p. 17. eBook ISBN 9781003178903. DOI: 10.1201/9781003178903-17.

Adetunji, C.O., Olaniyan, O.T., Adeyomoye, O., Dare, A., Adeniyi, M.J., Alex, E., Rebezov, M., Garipova, L., and Ali Shariati, M. (2022c). eHealth, mHealth, and Telemedicine for COVID-19 pandemic. In: Pani, S.K., Dash, S., dos Santos, W.P., Chan Bukhari, S.A., and Flammini, F. (eds.), *Assessing COVID-19 and Other Pandemics and Epidemics using Computational Modelling and Data Analysis*. Springer, Cham. https://doi.org/10.1007/978-3-030-79753-9_10

Adetunji, C.O., Olaniyan, O.T., Adeyomoye, O., Dare, A., Adeniyi, M.J., Alex, E., Rebezov, M., Petukhova, E., and Ali Shariati, M. (2022d). Machine learning approaches for COVID-19 pandemic. In: Pani, S.K., Dash, S., dos Santos, W.P., Chan Bukhari, S.A., and Flammini, F. (eds.). *Assessing COVID-19 and Other Pandemics and Epidemics using Computational Modelling and Data Analysis*. Springer, Cham. https://doi.org/10.1007/978-3-030-79753-9_8

Adetunji, C.O., Olaniyan, O.T., Adeyomoye, O., Dare, A., Adeniyi, M.J., Alex, E., Rebezov, M., Isabekova, O., and Ali Shariati, M. (2022e). Smart sensing for COVID-19 pandemic. In: Pani, S.K., Dash, S., dos Santos, W.P., Chan Bukhari, S.A., Flammini, F. (eds), *Assessing COVID-19 and Other Pandemics and Epidemics using Computational*

Modelling and Data Analysis. Springer, Cham. https://doi.org/10.1007/978-3-030-79753-9_9

Adetunji, C.O., Olaniyan, O.T., Adeyomoye, O., Dare, A., Adeniyi, M.J., Alex, E., Rebezov, M., Isabekova, O., and Ali Shariati, M. (2022f). Smart sensing for COVID-19 pandemic. In: Pani, S.K., Dash, S., dos Santos, W.P., Chan Bukhari, S.A., Flammini, F. (eds), *Assessing COVID-19 and Other Pandemics and Epidemics using Computational Modelling and Data Analysis*. Springer, Cham. https://doi.org/10.1007/978-3-030-79753-9_9

Adetunji, C.O., Olaniyan, O.T., Adeyomoye, O., Dare, A., Adeniyi, M.J., Alex, E., Rebezov, M., Koriagina, N., and Ali Shariati, M. (2022g). Diverse techniques applied for effective diagnosis of COVID-19. In: Pani S.K., Dash S., dos Santos W.P., Chan Bukhari S.A., and Flammini F. (eds.), *Assessing COVID-19 and Other Pandemics and Epidemics using Computational Modelling and Data Analysis*. Springer, Cham. https://doi.org/10.1007/978-3-030-79753-9_3

Adetunji, C. O., Oloke, J. K., Kumar, A., Swaranjit, S., and Akpor, B. (2017a). Synergetic effect of rhamnolipid from *Pseudomonas aeruginosa* C1501 and phytotoxic metabolite from *Lasiodiplodia pseudotheobromae* C1136 on *Amaranthus hybridus* L. and *Echinochloa crus-galli* weeds. *Environ. Sci. Pollut. Res.* 24 (15): 13700–13709. http://doi:10.1007/s11356-017-8983-8

Adetunji, C. O., Oloke, J. K., Pradeep, M., Jolly, R. S., Anil, K. S., Swaranjit, S. C., and Bello, O. M. (2017b) Characterization and optimization of a rhamnolipid from *Pseudomonas aeruginosa* C1501 with novel biosurfactant activities. *Sustain. Chem. Pharm.* 6: 26–36. http://dx.doi.org/10.1016/j.scp.2017.07.001

Adetunji, C.O., Olugbemi, O.T., Akram, M., Laila, U., Samuel, M.O., Oshinjo, A.M., Adetunji, J.B., Okotie, G.E., (Diuto) Esiobu, N., Oyedara, O.O., and Adeyemi, F.M. (2022h). Application of computational and bioinformatics techniques in drug repurposing for effective development of potential drug candidate for the management of COVID-19. In: *Medical Biotechnology, Biopharmaceutics, Forensic Science and Bioinformatics.* CRC Press, p.14. eBook ISBN 9781003178903. DOI: 10.1201/9781003178903-15

Adetunji, CO and Oyeyemi OT (2022i). Antiprotozoal activity of some medicinal plants against entamoeba histolytica, the causative agent of amoebiasis. In: *Medical Biotechnology, Biopharmaceutics, Forensic Science and Bioinformatics*. CRC Press, p. 12. eBook ISBN 9781003178903. www.taylorfrancis.com/chapters/edit/10.1201/9781003178903-20/antiprotozoal-activity-medicinal-plants-entamoeba-histolytica-causative-agent-amoebiasis-charles-oluwaseun-adetunji-oyetunde-oyeyemi.

Adetunji, C.O., and Ukhurebor, K.E. (2021). Recent trends in utilization of biotechnological tools for environmental sustainability. In: Adetunji, C.O., Panpatte, D.G., and Jhala, Y.K. (eds.), *Microbial Rejuvenation of Polluted Environment. Microorganisms for Sustainability*, vol. 27. Springer, Singapore. https://doi.org/10.1007/978-981-15-7459-7_11

Akram, M., Adetunji, C.O., Egbuna, C., Jabeen, S., Olaniyan, O., Ezeofor, N.J., Anani, O.A., Laila, U., Găman, M.-A., Patrick-Iwuanyanwu, K., Ifemeje, J.C., Chikwendu, C.J., Michael, O.C., and Rudrapal, M. 2021b. Dengue fever, Chapter 17. In: *Neglected Tropical Diseases and Phytochemicals in Drug Discovery*. Wiley. DOI: 10.1002/9781119617143

Akram, M., Mohiuddin, E., Adetunji, C.O., Oladosun, T.O., Ozolua, P., Olisaka, F.N., Egbuna, C., Michael, O., Adetunji, J.B., Hameed, L., Awuchi, C.G., Patrick-Iwuanyanwu, K., and Olaniyan, O. 2021a. Prospects of phytochemicals for the treatment of helminthiasis. In: *Neglected Tropical Diseases and Phytochemicals in Drug Discovery*. Chapter 7: Wiley. DOI: 10.1002/9781119617143

Ampatzidis, Y., Bellis, L. D., and Luvisi, A. (2017). iPathology: Robotic applications and management of plants and plant diseases. *Sustainability* 9(6): 1010. doi:10.3390/ su9061010

Aono, A. H., Costa, E. A., Rody, H. V. S., Nagai, J. S., Pimenta, R. J. G., Mancini, M. C.., dos Santos, F. R. C., Pinto, L. R., de Andrade Landell, M.G., de Souza, A. P., and Kuroshu, R. M. (2020). Machine learning approaches reveal genomic regions associated with sugarcane brown rust resistance. *Sci. Rep.* 10: 20057 https://doi.org/10.1038/s41598-020-77063-5

Araus, J. L., Kefauver, S. C., Zaman-Allah, M., Olsen, M.S., and Cairns, J. E. (2018). Translating high-throughput phenotyping into genetic gain. *Trends Plant Sci.* 23(5): 451–466. doi: 10.1016/j.tplants.2018.02.001

Arowora, K.A., Abiodun, A.A., Adetunji, C.O., Sanu, F.T., Afolayan, S.S., and Ogundele, B.A. (2012). Levels of aflatoxins in some agricultural commodities sold at Baboko Market in Ilorin, Nigeria. *Global J. Sci. Front. Res.* 12(10): 31–33.

Balamurugan S., Divyabharathi, N., Jayashruthi, K., Bowiya, M., Shermy, R., and Shanker, R. K. (2016) Internet of agriculture: Applying IoT to improve food and farming technology. *IRJET*. 3(10): 713–719. [Online]. Available: www.irjet.net/volume3-issue10.

Ballesteros, R., Ortega, J. F., Hernández, D., and Moreno, M. A. (2014). Applications of georeferenced high-resolution images obtained with unmanned aerial vehicles. Part II: application to maize and onion crops of a semi-arid region in Spain. *Precis. Agric.* 15: 593–614.

Bansod, B., Singh, R., Thakur, R., and Singhal, G. (2017). A comparison between satellite based and drone based remote sensing technology to achieve sustainable development: A review. *JAEID* .111(2): 383–407. DOI: 10.12895/jaeid.20172.690.

Barbedo, J. G. A. (2019). A review on the use of unmanned aerial vehicles and imaging sensors for monitoring and assessing plant stresses. *Drones*. 3: 40.

Bruno, S. F., Costa, F. G. Pessin, G., Ueyama, J., Freitas, H., Colombo, A., Fini, P. H., Villas, L. A., Osorio, F., Vargas, P. A., and Braun, T. (2014). The use of unmanned aerial vehicles and wireless sensor networks for spraying pesticides. *J. Syst. Archit.* 60(4): 393–404.

Cruzan, M. B., Weinstein, B. G., Grasty, M. R., Kohrn, B. F., Hendrickson, E. C., Arredondo, T. M., and Thompson, P. G. (2016). Small unmanned aerial vehicles (micro-UAVs, drones) in plant ecology. *Appl. Plant Sci.* 4(9). doi: 10.3732/apps.1600041

Dauda, W. P., Morumda, D., Abraham, P., Adetunji, C. O., Ghazanfar, S., Glen, E., Abraham, S. E., Peter, G. W., Ogra, I. O., Ifeanyi, U. J., Musa, H., Azameti, M. K., Paray, B. A., and Gulnaz, A. (2022a). Genome-wide analysis of cytochrome P450s of Alternaria species: Evolutionary Origin, family expansion and putative functions. *Journal of Fungi* 8(4): 324. https://doi.org/10.3390/jof8040324

Dauda, W.P., Abraham, P., Glen, E., Adetunji, C.O., Ghazanfar, S., Ali, S., Al-Zahrani, M., Azameti, M.K., Alao, S.E.L., Zarafi, A.B., Abraham, M.P., and Musa, H. (2022b). Robust profiling of cytochrome P450s (P450ome) in notable *Aspergillus* spp. *Life* 12(3): 451. https://doi.org/10.3390/life12030451

Debauche, O., Mahmoudi, S., Elmoulat, M., Mahmoudi, A.S., Manneback, P., and Lebeau, F.. (2020). Edge AI-IoT pivot irrigation, plant diseases, and pests identification. *Procedia Comput. Sci.*177: 40–48.

Del-Campo-Sanchez, A., Ballesteros, R., Hernandez-Lopez, D., Ortega, J. F., and Moreno, M. A. (2019). Quantifying the effect of Jacobiasca lybica pest on vineyards with UAVs by combining geometric and computer vision techniques. Agroforestry and Cartography Precision Research Group. *PLoS One*. 14(4): e0215521.

Diego, I. P. and Rieder, R. (2018). Computer vision and artificial intelligence in precision agriculture for grain crops: A systematic review. *Comput Electron Agric.* 153: 69–81.

Díez, C. (2017). Hacia una agricultura inteligente (Towards an intelligent agriculture). *Cuaderno de Campo*, 60: 4–11.

Dimitrios, P., and Konstantinos, A. (2016). SensoTube: A scalable hardware design architecture for wireless sensors and actuators. *Networks Nodes in the Agricultural Domain. Sensors*. 16:1227–1246.

Durgabai, R. P. L., Bhargavi, P. and Jyothi, S. (2018) Pest Management Using Machine Learning Algorithms: A Review. *IJCSEITR*. 8(1): 13–22. ISSN (P): 2249-6831; ISSN (E): 2249–7943.

Elijah, O., Rahman, T. A., Orikumhi, I., Leow, C. Y., and Hindia, M. N. (2018). An overview of Internet of Things (IoT) and data analytics in agriculture: Benefits and challenges. *IEEE Internet Things J*. 5:3758–3773. doi: 10.1109/JIOT.2018.2844296

Everitt, J. H., Summy, K. R., Escobar, D. E., and Davis, M. R. (2003). An overview of aircraft remote sensing in integrated pest management. *Subtropical Plant Science*, 55: 59–67.

Faical, B., Freitas, P. H.. Gomes, L. Y., Mano, G. Pessin, A. C. P. and Ueyama, J. 2017. An adaptive approach for UAV based pesticide spraying in dynamic environments. *Comput. Electron. Agric.* 138: 210–223.

Fedor, Peter, Jaromi´ R. Van Hara, Josef Havel, Igor Malenovsky and Ian Spellerberg (2009) Artificial intelligence in pest insect monitoring. *Syst. Entomol.*, 34: 398–400 DOI: 10.1111/j.1365-3113.2008.00461.x

Fernandez-Quintanilla, C., Pena, J., Andujar, D., Dorado, J., Ribeiro, A., and Lopez-Granados, F. (2018). Is the current state of the art of weed monitoring suitable for site specific weed? *Weed Research International Journal of Weed Biology, Ecology and Vegetation Management*, 58(4): 259–272.

Filho, F. H. (2019). *Remote Sensing for Monitoring Whitefly, Bemisia Tabaci Biotype B (Hemiptera: Aleyrodidae) in Soybean.* Master's Thesis. University of São Paulo Piracicaba, São Paulo, Brazil (in Portuguese with English abstract).

Filho, F. H., Heldens, W.B., Kong, Z., and de Lange, E.S. (2019). Drones: Innovative technology for use in precision pest management. *J. Econ. Entomol.*, 113(1): 1–25.

Filho, F. H., Heldens, W.B., Kong, Z., and de Lange, E.S. (2020). Drones: Innovative technology for use in precision pest management. *J. Econ. Entomol.*, 113(1): 1–25, https://doi.org/10.1093/jee/toz268

Gao, D., Quan, S., Hu, B., and Zhang, S. (2020). A framework for agricultural pest and disease monitoring based on internet-of-things and unmanned aerial vehicles. *Sensors (Basel).* 20(5): 1487. Doi: 10.3390/s20051487

Golhani. K., Balasundram, S. K., Vadamalai, G., and Pradhan, B. (2018). A review of neural networks in plant disease detection using hyperspectral data. *Inf. Process. Agric*. 5: 354–371. Doi: 10.1016/j.inpa.2018.05.002

Guijarro, M., Pajares, G., Riomoros, I., Herrera, P. J., Burgos-Artizzu, X. P., and Ribeiro, A. (2011). Automatic segmentation of relevant textures in agricultural images. *Comput. Electron. Agric*. 75: 75–83.

Hassan, A., Juraimi, A.S., and Saiful, M. H. (2016). Introduction to robotics agriculture in pest control: A review. *PJSRR* 2(2): 80–93 eISSN: 2462-2028 © Universiti Putra Malaysia Press.

Himesh, S. (2018). Digital revolution and big data: A new revolution in agriculture. *CAB Rev*. 13, 1–7.

Huang, X., Xiao, J., and Ma, M. (2019). Evaluating the performance of satellite-derived vegetation indices for estimating gross primary productivity using FLUXNET observations across the globe. *Remote Sens*. 11(15): 1823. https://doi.org/10.3390/rs11151823

Joalland, S., Screpanti, C., Varella, H. V, Reuther, M., Schwind, M., Lang C., et al. (2018). Aerial and ground based sensing of tolerance to beet cyst nematode in sugar beet. *Remote Sens*. 10: 787.

Judd, G., and Gardiner, M. G. T. (2005). Towards eradication of codling moth in British Columbia by complimentary actions of mating disruption, tree banding and sterile insect technique: five year study in organic orchards. *Crop Prot.* 24: 718–733.

Lee, H., Moon, A., Moon, K., and Lee, Y. (2017). Disease and pest prediction IoT system in orchard: A preliminary study. *Proceedings of the 2017 Ninth International Conference on Ubiquitous and Future Networks (ICUFN)*; Milan, Italy. 4–7 July; pp. 525–527.

Li, X., Pak, C., and Bi, K. (2020). Analysis of the development trends and innovation characteristics of Internet of Things technology-based on patentometrics and bibliometrics. *Technol. Anal. Strateg. Manag*. 32:104–118. doi: 10.1080/09537325.2019.1636960

López, J. J. and Mulero-Pázmány, M. (2019). Drones for conservation in protected areas: Present and future. *Drones*. 3(1), 10; https://doi.org/10.3390/drones3010010.

M'mboyi, F., Mugo, S., Murenga, M., and Ambani, L. (2010). Maize production and improvement in Sub-Saharan Africa. In *Nairobi, Kenya: African Biotechnology Stakeholders Forum (ABSF). management in arable crops? Weed Research* 58(4):259–272.

Ma, J., and Zhang X. Y. (2018). Application of Internet of Things Technology in Monitoring and Early Warning of Diseases and Pests in Ningxia. *Forest Investig. Plan*. 43:71–74.

Mattupalli, C., Komp, M. R., and Young, C. A. (2017). Integrating geospatial technologies and unmanned aircraft systems into the grower's disease management toolbox. *APS Features*. doi:10.1094/APSFeature-2017-7

Meshesha, D. T., Muhyadin Mohammed Ahmed, Dahir Yosuf Abdi, and Nigussie Haregeweync (2020). Prediction of grass biomass from satellite imagery in Somali regional state, eastern Ethiopia. *Heliyon*. 2020 Oct; 6(10): doi: 10.1016/j.heliyon.2020.e05272

Michelmore, R., Coaker, G., Bart, R., Beattie, G., Bent, A., Bruce, T., Cameron, D., Dangl, J., Dinesh-Kumar, S., Edwards, R., Eves-van den Akker, S., Gassmann, W., Greenberg, J. T., Hanley-Bowdoin, L., Harrison, R. J., Harvey, J., He, P., Huffaker, A., Hulbert, S., Innes, R., Jones. J. D. G., Kaloshian, I., Kamoun, S., Katagiri, F., Leach, J., Ma, W., McDowell, J., Medford, J., Meyers, B., Nelson, R., Oliver, R., Qi, Y., Saunders, D., Shaw, M., Smart, C., Subudhi, P., Torrance, L., Tyler, B., Valent, B., and Walsh, J. (2017). Foundational and translational research opportunities to improve plant health. *Mol. Plant Microbe. Interact*. 30(7): 515–516. doi: 10.1094/MPMI-01-17-0010-CR. Epub 2017 Jun 12. PMID: 28398839; PMCID: PMC5810936.

Ngugi, L. C., Abelwahab, M., and Abo-Zahhad, M. (2020). Recent advances in image processing techniques for automated leaf pest and disease recognition – A review. *Information Processing in Agriculture*. https://doi.org/10.1016/j.inpa.2020.04.004.

Nwankwo, W., Adetunji, C.O., Ukhurebor, K.E., Panpatte, D.G., Makinde, A.S., and Hefft, D.I. (2021). Recent Advances in Application of Microbial Enzymes for Biodegradation of Waste and Hazardous Waste Material. In: Adetunji, C.O., Panpatte, D.G., Jhala, Y.K. (eds), Microbial Rejuvenation of Polluted Environment. *Microorganisms for Sustainability*, vol 27. Springer, Singapore. https://doi.org/10.1007/978-981-15-7459-7_3

Okeke, N.E., Adetunji, C.O., Nwankwo, W., Ukhurebor, K.E., Makinde, A.S., Panpatte, D.G. (2021). A Critical Review of Microbial Transport in Effluent Waste and Sewage Sludge Treatment. In: Adetunji, C.O., Panpatte, D.G., Jhala, Y.K. (eds), *Microbial Rejuvenation of Polluted Environment. Microorganisms for Sustainability*, vol 27. Springer, Singapore. https://doi.org/10.1007/978-981-15-7459-7_10

Olaniyan, O. T., Adetunji, C.O., Adeniyi, M. J., and Hefft, D. I. (2022). Computational intelligence in IoT healthcare. 2022. In: *Deep Learning, Machine Learning and IoT in*

Biomedical and Health Informatics. CRC Press, p. 13. eBook ISBN 9780367548445. DOI: 10.1201/9780367548445-19

Olaniyan, O.T., Adetunji, C.O., Adeniyi, M.J., and Hefft, D. H. (2022). Machine learning techniques for high-performance computing for IoT applications in healthcare. In: *Deep Learning, Machine Learning and IoT in Biomedical and Health Informatics*. CRC Press, p. 13. eBook ISBN 9780367548445. DOI: 10.1201/9780367548445-20.

Oluwaseun, A.C., Phazang, P., and Sarin, N.B. (2017). Significance of rhamnolipids as a biological control agent in the management of crops/plant pathogens, *Current Trends in Biomedical Engineering & Biosciences*, Juniper Publishers Inc., 10(3): 54–55, December.

Owomugisha, G., and E. Mwebaze (2016). Machine learning for plant disease incidence and severity measurements from leaf images, *15th IEEE International Conference on Machine Learning and Applications* (ICMLA), pp. 158–163.

Oyedara, O.O., Folasade Muibat Adeyemi, Charles Oluwaseun Adetunji, Temidayo Oluyomi Elufisan. (2022). Repositioning antiviral drugs as a rapid and cost-effective approach to discover treatment against SARS-CoV-2 infection. In: *Medical Biotechnology, Biopharmaceutics, Forensic Science and Bioinformatics*. CRC Press, p. 12. eBook ISBN 9781003178903. DOI: 10.1201/9781003178903-10.

Panda, C.K. (2019). Advances in application of ICT in crop. Pest and disease management. In: Egbuna, C., and Sawicka, B. (eds.), *Natural Remedies for Pest, Disease and Weed Control*. Academic Press, pp. 235–242.

Patil, K., Pattewar, P., and Nandgave, S. (2020). Disease detection application for crops using augmented reality and artificial intelligence. *IRJET*, 7 (3): 4089–4092. e-ISSN: 2395-0056

Patil, N., Khope, B., Akshata, N., Lokesha, H., Shariff, M., and Kha, M. (2019). The economics of applications of artificial intelligence and machine learning in agriculture. *Int. J. Pure Appl. Biosci.*7 (1): 296–305. DOI: http://dx.doi.org/10.18782/2320-7051.7324

Pettorelli, N., Vik, J.O., Mysterud, A., Gaillard, J.-M., Tucker, C.J., and Chr. Stenseth, N. (2005). Using the satellite-derived NDVI to assess ecological responses to environmental change. *Trends Ecol. Evol.* 20(9): 503–510. DOI: https://doi.org/10.1016/j.tree.2005.05.011

Pretty, J., and Bharucha, Z.P. (2015). Integrated pest management for sustainable intensification of agriculture in Asia and Africa. *Insects*, 6: 152–182. Doi:10.3390/insects6010152

Puente, M., Darnall, N., and Forkner, R.E. (2011). Assessing integrated pest management adoption: Measurement problems and policy implications. *Environ. Manag.* 48: 1013–1023. DOI 10.1007/s00267-011-9737-x

Puig, E., Gonzalez, F., Hamilton, G., and Grundy, P. (2015). Assessment of crop insect damage using unmanned aerial systems: A machine learning approach. *21st International Congress on Modelling and Simulation*. Gold Coast, Australia, pp. 1420–1426. www.mssanz.org.au/modsim2015.

Redmond, R.S., Weltzien, C., Hameed, I.A., Yule, I.J. Grift, T.E., Balasundram, S.K., Pitonakova, L., Ahmad, D., and Chowdhary, G. (2018). Research and development in agricultural robotics: A perspective of digital farming. *Int. J. Agric. & Biol. Eng.* 11(4). Open Access at www.ijabe.org

Rudic, T., Lindsay, E., McCulloch, A., and Cushman, K. (2020). Comparison of smartphone and drone Lidar methods for characterizing spatial variation in PAI in a tropical forest. *Remote Sens.* 12(11): 1765. https://doi.org/10.3390/rs12111765

Rumpf, T., Mahlein, A.-K., Steiner, U., Oerke, E.-C., Dehne, H.W., and Plumer. 2020. Early detection and classification of plant diseases with support vector machines based on hyperspectral reflectance. *Comput Electron Agric* . 74(1): 91–99.

Saiz-Rubio, V. and Rovira-Más, F. (2020). From: Smart farming towards Agriculture 5.0: A review on crop data management. Agricultural Robotics Laboratory (ARL), Universitat Politè. *Agronomy*, 10, 207: 1–21. doi:10.3390/agronomy10020207

Sangeetha, J., Hospet, R., Thangadurai, D., Adetunji, C.O., Islam, S., Pujari, N., and Al-Tawaha, A.R.M.S. (2021). Nanopesticides, nanoherbicides, and nanofertilizers: The greener aspects of agrochemical synthesis using nanotools and nanoprocesses toward sustainable agriculture. In: Kharissova, O.V., Torres-Martínez, L.M., and Kharisov, B.I. (eds.), *Handbook of Nanomaterials and Nanocomposites for Energy and Environmental Applications*. Springer, Cham. https://doi.org/10.1007/978-3-030-36268-3_44

Sanghvi, Y., Gupta, H., Doshi, H., Koli, D., Ansh, A., and Gupta, U. (2015). Comparison of self-organizing maps and Sammon's mapping on agricultural datasets for precision agriculture. *Proceedings of the 2015 International Conference on Innovations in Information, Embedded and Communication Systems (ICIIECS)*, Coimbatore, India, 19–20 March, pp. 1–5.

Sankaran, S., Mishra, A., Ehsani, R., and Davis, C. (2010). A review of advanced techniques for detecting plant diseases. *Comput Electron Agric*. 72:1–13. Doi: 10.1016/j.compag.2010.02.007

Schimmelpfennig, D. (2016). Farm profits and adoption of precision agriculture. *USDA*, 217, 1–46.

Selvaraj, M.G., Vergara, A., Ruiz, H., Safari, N., Elayabalan, S., Ocimati, W., and Blomme, G. (2019). AI-powered banana diseases and pest detection. *Plant Methods*, 15(92). https://doi.org/10.1186/s13007-019-0475-z.

Shafi, U., Mumtaz, R., García-Nieto, J., Ali Hassan, S., Ali Raza Zaidi, S., and Iqbal, N. (2019). Precision agriculture techniques and practices: From considerations to applications. *Sensors (Basel)*. 19(17): 3796. Published online 2019 Sep 2. Doi: 10.3390/s19173796.

Shamshiri, R.R., Hameed, I.A, Karkee, M., and Weltzien, C. (2018). Robotic harvesting of fruiting vegetables: A simulation approach in V-REP, ROS and MATLAB. *Proceedings in Automation in Agriculture-Securing Food Supplies for Future Generations. InTech.*

Shamshiri, R.R., Weltzien, C., Hameed, I.A., Yule, I.J., Grift, T.E., Balasundram, S.K., et al. (2018). Research and development in agricultural robotics: A perspective of digital farming. *Int. J. Agric. Biol.*, 11(4): 1–14.

Singh, A., Dhiman, N., Kar Kumar, A., Singh, D., Purohit, P.M., Ghosh, D., and Patnaik, S. (2019). Advances in controlled release of pesticide formulations: Prospect to safer integrated pest management and sustainable agriculture. *J. Hazard. Mater.* 385:121525.

Sreelakshmi, M., and Padmanayana (2015). Early detection and classification of pests using image processing. *International Journal of Innovative Research in Electrical, Electronics, Instrumentation and Control Engineering and National Conference on Advanced Innovation in Engineering and Technology (NCAIET-2015) Alva's Institute of Engineering and Technology, Moodbidri.* Vol. 3, Special Issue 1.

Stanton, C., Starek, M.J., Elliott, N., Brewer, M., Maeda, M.M., and Chu, T. (2017). Unmanned aircraft system-derived crop height and normalized difference vegetation index metrics for sorghum yield and aphid stress assessment. *J. Appl. Remote Sens.* 11(2): 026035, doi: 10.1117/1.JRS.11.026035

Strickland, M. R., Ess, D. R., and Parsons, S. D. (1998). Precision farming and precision pest management: the power of new crop production technologies. *J. Nematol.* 30(4):431–435.

Suresh, K.P., Dhemadri, Kurli, R., Dheeraj, R. and Roy, P. (2019). Application of Artificial Intelligence for livestock disease prediction. *Indian Farming*, 69(03): 60–62.

Ukhurebor, K.E., Adetunji, C.O., Bobadoye, A.O., Aigbe, U.O., Onyancha, R.B., Siloko, I.U., Emegha, J.O., Okocha, G.O., and Abiodun, I.C. (2021). Bionanomaterials for Biosensor Technology. *Bionanomaterials. Fundamentals and biomedical applications*, pp. 5–22.

Ukhurebor, K.E., Mishra, P., Mishra, R.R., and Adetunji, C.O. (2020). Nexus between climate change and food innovation technology: Recent advances. In: Mishra, P., Mishra, R.R., and Adetunji, C.O. (eds.), *Innovations in Food Technology*. Springer, Singapore. https://doi.org/10.1007/978-981-15-6121-4_20

Ukhurebor, K.E., Nwankwo, W., Adetunji, C.O., and Makinde, A.S. (2021). Artificial intelligence and internet of things in instrumentation and control in waste biodegradation plants: recent developments. In: Adetunji, C.O., Panpatte, D.G., and Jhala, Y.K. (eds.), *Microbial Rejuvenation of Polluted Environment. Microorganisms for Sustainability*, vol. 27. Springer, Singapore. https://doi.org/10.1007/978-981-15-7459-7_12

Vescovo, L., Wohl Fahrt, G., Balzarolo, M., Pilloni, S., Sottocornola, M., Rodeghiero, M., and Gianelle, D. (2012). New spectral vegetation indices based on the near-infrared shoulder wavelengths for remote detection of grassland phytomass. *Int. J. Remote Sens.* 33(7). doi: 10.1080/01431161.2011.607195

Vreysen, M.J.B., Robinson, A.S., and Hendrichs, J. (eds.) (2007). Area-wide control of insect pests: From research to field implementation. *IAEA*, 3–33.

Weiss, U. and Biber, P. (2011). Plant detection and mapping for agricultural robots using a 3D LIDAR sensor. *Rob Auton Syst.* 59(5): 265–273. doi:10.1016/j.robot.2011.02.011

Zhang, Y. (2019). The role of precision agriculture. *Resource*, 19, 9.

5 The Role of Biosensors in Inspection of Food and Agricultural Products

Current Advances

Charles Oluwaseun Adetunji, Abel Inobeme, Inamuddin, John Tsado Mathew, Jonathan Inobeme, Alexander Ikechukwu Ajai, Juliana Bunmi Adetunji, Maliki Munirat, Akinola Samson Olayinka, and Olalekan Akinbo

CONTENTS

5.1 INTRODUCTION

It has been observed that microbial diseases, most especially infections due to bacteria, have the capability to spread easily. This might be linked to the increase in the trade of agricultural products worldwide. Numerous microorganisms have been recognized to be responsible for several health challenges. It has been forecasted that approximately 2 million deaths are recorded annually, which might be linked to unsafe food, and the level of mortality recorded annually could be responsible for many diseases, varying from common diarrhea to cases of cancers and tumor growth. Hence, the World

DOI: 10.1201/9781003207955-5

Health Organization (WHO) has encouraged food safety. (WHO, 2015). It has also been stated by the United State Department of Agriculture that the costs of foodborne diseases could amount about $15.6 billion each year (CDC, 2016). Typical examples of these foodborne diseases include *Listeria Monocytogenes* and *Salmonella spp.*, amongst others (Alocilja and Radke, 2003). Moreover, it has been discovered that foodborne illness could be linked to some pathogenic microorganisms as a causative agent for various diseases such as blindness, intestinal inflammation, reactive arthritis, chronic kidney diseases, and mental disability (Hoffmann et al., 2015).

One of the approaches in the classification of biosensors is according to their principle of transduction. The various types of biosensors include optical, piezoelectric (such as quarz crystal type), and electrochemical or, by their feature of identification, as enzymatic sensors, genosensors, and aptasensors depending on the biological component that is employed. It has been discovered that biosensors have a wider application in detection of pollutants available in the food samples or in the environment (Adetunji et al., 2021a,b,c; Ukhurebor et al., 2021; Sangeetha et al., 2021; Adetunji and Ukhurebor, 2021, Oluwaseun et al., 2017; Adetunji et al., 2012; Arowora et al., 2012; Dauda et al., 2022a,b; Ukhurebor et al. 2021; Okeke et al., 2021; Adetunji and Anani, 2021; Nwankwo et al., 2021; Adejumo and Adetunji, 2018; Ukhurebor et al. 2021). Moreover, most of biosensors are known as enzyme based or immunosensor type, but the production of aptasensors has recently been increased drastically due to several benefits, which include ease of utilization and modification, high resistance to heat, in vitro preparation and the possibility of structuring them as well as distinguishing targeted pollutants from different functional groups and rehybridizing them (Adetunji et al., 2020; Adetunji et al., 2020a,b,c; Oyedara et al., 2022; Adetunji et al., 2022a,b,c,d,e,f,g,h,i; Olaniyan et al., 2022).

It has been discovered that the conventional methods ofrecisionon of pollutant involves modern analytical techniques such as different chromatographic techniques (e.g. liquid chromatography and high-pressure chromatography combined with celectrophoresic or mass spectrometric method), which include costly chemicals, sample preparation procedure that takes time and instruments that are not easily affordable. This has necessitate a high demand for a more sensitive, reliable, fast, cost effective biosensor that could detect these contaminants and foodborne related pathogens.

Moreover, the little success have been recorded in through the application of traditional methods most especially during the in situ measurements such as sceneriors where food toxins, pesticides, are released unknowingly. Hence, there is a need for more precise, sensitive portable equipment such as biosensors for effective determination of these pollutants (Arduini et al., 2013).

With respect to this, the position of nanotechnology in the production of rapid and advanced biosensing instruments is fundamental to the successful detection of pollutants. Also, it has been discovered that the majority of recent biosensors are made up of nanomaterials, and new nanocomposites in their system, possessing numerous advantages in improving analytical effectiveness, which includes better sensitivity and detection limitations (Maduraiveeran and Jin, 2017). This might be linked to their small sizes and strong capacity for mediation of electrons. Also, the application of nanoparticles from gold could also be a hopeful and flexible platform for enzyme immobilization matrix, which could ensure a better stabilization of the enzyme through electrostatic interactions.

Furthermore, apart from the better efficiency that has been displayed by gold nanoparticles for exceptional biocompatibility and minute cytotoxicity effects when performed during in vivo assay. The inherent role of nanotechnology alongside biotechnology may boost the diagnostic performance of sensing devices with a view to developing more viable biosensors for in situ detection, which could mitigate all the aforementioned challenges related to conventional analytical tools.

Moreover, there are several challenges encountered as major obstacles in the marketing of these biosensors, which might be linked to the interdisciplinary manufacturing background and limitations on in situ activity, during analytical performance, and primarily in reproducibility; (Guo et al., 2017). Also, the biosensors that are available in the markets are also limited to their field applications due to the fact that most biosensors are successfully tested in buffered systems or purified water when exposed to pollutants present in real samples. The presence of these pollutants might affect their analytical output majorly by the matrix impact (Bahadir and Sezgintürk, 2015).

Some typical examples of biosensors that could be employed for monitoring of pollutant in food materials includes optical immunosensors based on the 2,4-dichlorophenol surface plasmon resonance principle and dichlorodiphenyltrichloroethane (DDT) (Soh et al., 2003) Furthermore, some other examples include their application in detecting disease-causing microorganisms such as fungi and bacteria as well as in identifying the presence of pesticide residues such as organophosphates. They can also analyze the existence of highly toxic metals such as mercury ions as well as other chemicals that can bring about the disruption of the human endocrine system.

Therefore, this chapter provides detailed information on the application of biosensors in the determination of contaminants and foodborne microbes. Recent advances and improvements in technology that have led to increases in sensitivity of the instrument for contaminants, food poisoning detection are also highlighted.

5.2 GENERAL OVERVIEW ABOUT THE APPLICATION OF TOXIC GAS SENSORS AND BIOSENSORS AND THEIR APPLICATION IN FOOD INSPECTION

Surveillance of food protection and nutritional quality is very important. The traditional analytical methods for quality analysis are very complex, consume time, and require qualified personnel – hence, swift, responsive and accurate methods need to be established for fast assessment of food quality in relation to human health (Thakur and Ragavan, 2013). Consumers also expect reliable food quality at a reasonable amount, long storage potential, and excellent product physical condition, while food analyst and assessors demand healthy processing practices, effective packaging as well as enforcement (Takhistov, 2005).

Numerous efforts to address problems and improve the drivers in food processing have been promoted. Biosensing technologies and their remarkable applications are extensively used to tackle top food production challenges and their sustainability. In this regard it is necessary to ensure proper packaging of food for efficient preservation and safety. Among the packaging optimization techniques for minimizing food waste is size diversification to help customers purchase the right amount and innovative packaging designs to prevent fragrance loss and offer sufficient moisture content (Jiang et al., 2016). There are two ways that intelligent packaging systems could help in data systems employed in storing or transmitting information and incident or biosensor indicators in packaging that allow environmental protection and product packaging. The interest in food quality and safety needs the production of responsive and effective analytical techniques, as well as freshness preservation and food quality technology (Mustafa and Silvana Andreescu, 2018).

Due to their special chemical and electrooptical properties, nanosensors can strengthen the drawbacks associated with the packaging of food. They are capable of detecting the presence of contaminants and even environmental shifts. Nanosensors help to ensure customers buy good items with appreciable taste, and the incidence of foodborne diseases that threaten food safety is reduced (Omanović-Mikličanina and Maksimović, 2016). Sensor systems provide significant advantages when compared to the existing systems, since they are cheap and portable, but they must be incorporated as a process into existing processing systems in analytical control (PAT) devices. Quality assurance in the production of food is related to the product's technology and sensory characteristics, microbiological health, chemical content, and nutritional value Biosensor instruments are emergent as among the most important detecting methods for general evaluation because of their speedrecisionn, and ease of production, as well as their applicability to diverse fields. Consequently, food sustainability is experiencing a increasing demand in biosensing technologies. Microfluidics is a scientific system that integrates various techniques (Adley, 2014).

With respect to biosensing technology, nanomaterials are considered the best tool in solving wellbeing, energy, and ecological concerns methodically linked to world populaces. The need for point-of-care technique with respect to this is focused on analytical instruments that are fast, simple, reliable, compact, and lower cost. A specificity arises from a variety of interactions involving antigens, peptide, cofactor, chemical processes, and recombination of nucleic acid associated with several transducers (Thakur and Ragavan, 2013. Numerous nanosensors are constructed for different uses in the agriculture and food industries to either rapidly diagnose threats, in the case of potential food poisoning, or to be incorporated into labeling as nanotracers to reveal the food product history and whether it is of good quality at a given time. There is a huge request for new, swift, and resourceful identification of constituents of food and beverages, including highly specific foodborne and waterborne microbes, contaminants, and traces of pesticides (Pividori and Alegret, 2010).

Biosensors have also been described as effective, powerful alternative techniques because of their efficient dissemination, quick and high-reliability performance. Biosensors have broader applications in the food and bioprocessing sectors in different countries to ensure nutritional quality and adequate consumer safety. An efficient food-safety strategy needs to be established to track the threats to human health linked to pollutants in raw resources, agricultural systems, and food production techniques. Contaminants in food might include microbiological (bacteria, viruses, pathogens), exogenous whether biological or physical, organic chemical compounds (pesticides, heavy metals, agricultural contaminants, undesirable fermentation products), and packaging materials.

Few of the agents in food are eco-friendly natural contaminants, but others are intentionally added compounds during food manufacturing (Pividori and Alegret, 2010). Salmonella is among the most common foodborne microbes for animal ailments in milk that affect the microbial quality of foodstuffs. Government bodies responsible for monitoring food safety, like those in the United States, propose methodologies of conventional culture to remove salmonella from meat and other food products

Developing new techniques with uniqueness in terms of quick response, flexibility, and convenience of multiplexing is a major concern for food safety inspection aimed at detecting negative specimens. Electrochemical magnetic sensing electrodes are a fast and simple way of detecting salmonella in milk. The bacteria are collected from milk specimens with magnetic beads via this process, and are preconcentrated by an immunological process with a strong immunity against *Salmonella spp.*

The next polyclonal peroxidase-labeled antibody, with magneto-electrode-based electrochemical detection, is used as serological confirmation. Better performance has been obtained through a single-step immunological process among the different procedures (Pividori and Alegret, 2010). It is regrettable that, in spite of strict regulations put forward in various countries, some foods that pose a hazard to humans still find their way into markets. Enzymatic and potentiometric biosensors have also been very useful in this regard in monitoring food items, especially in beer, to check aging (Mustafa and Andreescu, 2018). Microbial presence such as E. coli and other toxic organisms can also be checked effectively through biosensing devices. Some organisms that are thermophilic in nature are usually found in milk and related products can also be checked through biosensors. Furthermore, the incorporation of micro- and nanostructured materials into biosensing tools has provided good analytical efficacy in the identification of food contaminants (mostly antibiotics and pesticide residues), chemical pollutants, and irritants, and microbes commonly found in food (Lorenzo et al., 2018).

One of the main advances in the area of electrochemical biosensing is the innovation of new transducers, not just with better transductive parameters as well as better immobilization of biomolecules, while maintaining their metabolic activity. The incorporation of magnetic particles further improves the identification of food contaminants in a complicated matrix, thus enhancing the electrochemical signals due to the bigger size of the active transducer region Such devices have been applied to new ideas and revolutionary prototypes relating to food inspection, safety, and preservation for novel separators, electrodes, and catalysts (Pividori and Alegret, 2010).

Foodomics, or food fingerprinting, deals with nutritional standards, consistency and validity, and with food health and protection (Gallo and Ferranti, 2016). Integrated analytical technologies that rely on innovative food platforms can be used to identify food fingerprints include improved analytical procedures, phytochemistry, food chemistry, and bioinformatics, and biosensors include improved analytical techniques, food chemistry, and bioinformatics. Security and sustainability in relation to globalized changes in the environment are vital in this regard (Jiang et al., 2016).

It is expected that powerful analytical approaches will discover novel biomarkers, ensuring food safety and food security and encouraging individual peculiarities and personalized prognoses in agricultural production (Pividori and Alegret, 2010; Neethirajan et al., 2018. Multiple biosensors for assessing and controlling food quality have been developed. For example, a biosensor was developed to detect toxic chemicals based on water-soluble biocompatible oligoaziridine with pendent anthracene units and an on-off or off-off characteristic. Extremely accurate latest technology-remote sensing was created using laser range and laser scanning in crop height and vitality management.

One of the main possible targets for bioterrorism is the food industry. Much of the damage is caused by (1) end product contamination with chemical and biological agents designed to cause disease in customers; (2) interruption of diet delivery schemes; (3) harm to the agricultural cycle by the introduction of harmful crop pests diseases, which can have a significant effect on the food system.

For a number of reasons, attempts to establish awareness of preparedness and response measures to protect the country's food sources face significant challenges. The food system includes many different sectors; a wide array of biochemical agents may eventually pollute the supply of food, and the possible consequences for intentional contamination are practically infinite. The healthcare system is dynamic, and the responsibilities in the hindrance of foodborne illnesses can overlap in the gray area, or even worse, can fail. The key issue that the food safety plan will tackle is the timely detection of hazardous items, including leadership for the strategy and implementation of such a plan into the accessible supervision tools for product safety. The protective recognition of health hazards can be proficient by determination with the biosensors, or through the analysis and regulation of the process as well as the environment. The discovery is established on data from chemical as well as physical sensors with robust and permit scale-down, which indicates that the current information carriers can be easily integrated (Takhistov, 2005).

Biosensors may also be used in foodstuffs, medicines and manufacturing industries to detect pathogens and other microbes. Significant progress and development have been witnessed in this region through micro-level research and engineering biology interfaces with novel materials, including nanomaterials (Saleem, 2013).

The part of food processing where there is the most obvious need for the nanosensors is food processing and food transport. Product packaging removes sensory input from product, so consumers must rely on manufacturers' "expiry dates" based on a set of idealized expectations about how the product is processed or transported. If the conditions of transport are violated for some period of time, the quality of the food may deteriorate and may not be known to the customer until after opening the food package (Joyner and Kumar, 2015).

Food protection authorities have built strict surveillance programs designed to prevent chemicals from coming in contact with the supply of food, so official research labs have to be capable of processing a large number of samples reliably. There is indeed an imperative necessity to produce quick, economical, responsive, extraordinary sample throughput and possibility to employ quickly in the field as an "alarm" to easily determine the menace of poisoning by food microbes in a extensive range of food milieus, notably because existing techniques do not meet these detection criteria. The superior sensitivity of nanosensors and aptamers enables the detection of different of analytes, such as metal ions, contaminants, viruses, biomolecules, nucleic acids, and proteins. Many platforms were designed to rejoin to gas generation throughout maturing progressions in the headspace of packaging. (Esser, et al., 2012).

Many systems require monitoring systems through identification of radio frequencies (RFID). An RFID sensor has been produced by researchers to identify *Escherichia spp.* and *Salmonella spp* in processed foods. This research focuses on immobilized antitoxins based on the embedded versatile RFID tags. Basically, the system is linked to a wireless network, and the farmers and suppliers are providing a visual alert to track in real time.

Another framework was developed in New Zealand as a smart ripeness-indicator mark: RipeSense (RipeSense, Auckland) (Jones et al., 2005). The system is capable of conveying the degree of fruit ripeness without any need to break the package, but only by examining the difference in the color of the label that interacts with the gasses placed on top of the fruit packaging. Ethylene is the most commonly used predictor of maturation emitted during the maturation process among the evolved gasses (Mustafa and Andreescu, 2018).

Figure 5.1 shows areas of sensors' application in food safety and control.

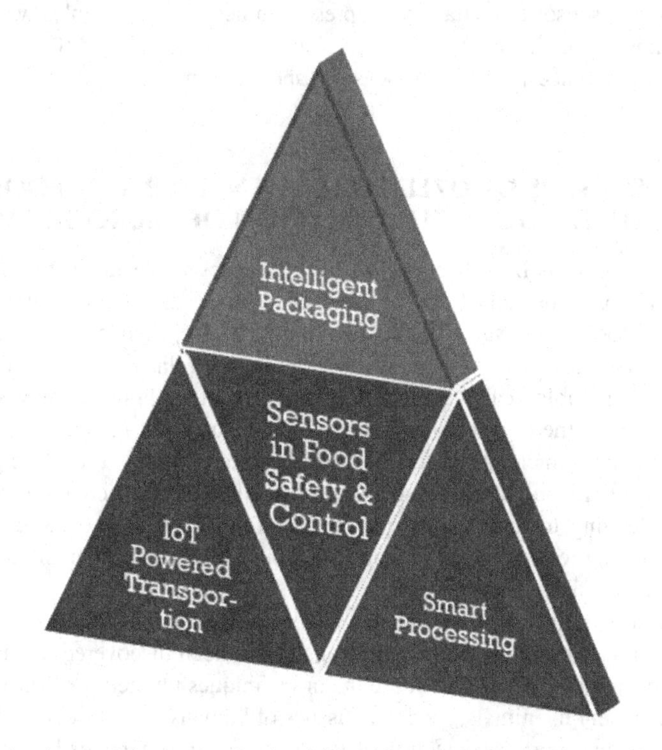

FIGURE 5.1 Areas of Sensor Application in Food Safety and Control.

5.3 E-NOSE TECHNOLOGY

Another important field is where biosensors have made a significant contribution to food inspection and safety. Eggs stored for varying periods of time and kept at chilled or at room temperature may be differentiated by an "e-nose." An e-nose based on ion-mobility was used to assess the distinction of hard and extra hard cheese samples, as well as age-dependent discrimination of cheeses or origin (Ampuero and Bosset, 2003). During the aging cycle, the e-nose was used in meat to detect bacterial spoilage using biosensors that included a silver or platinum electrode on which the enzymes putrescine or xanthin oxidases were immobilized (Priyanka, S Bhupinder, 2016).

Fish odor is a significant consistency parameter that is acceptable or not on the basis of performance of fish and fish products, normally achieved by sensory or gas chromatography. Therefore, an efficient technique for controlling the quality of fish and fishery products is required. Electronic noses play an important role by providing fast, automated and objective tools for quality control of fish. Also, freshness in fish has been calculated by measuring the related volatile compounds consisting of alcohols, carbonyls, amines, and meraptans that have shown typical changes in concentration over time under different storage conditions (Handa and Singh, 2018).

The e-nose is used to differentiate samples of beer and also highlights the compound that makes the significant variations. The sensor-based electronic noses are used to identify efficient technology for the development of various beer types. The ease of the aptasensor to contaminants present in actual field samples was achieved by examining shellfish and by obtaining reliable recoveries (102–110 percent), indicating no interference with the aptasensor response from the shellfish matrix (Wang et al., 2015).

5.4 BIOSENSORS UTILIZED FOR ENHANCEMENT OF FOOD SAFETY THROUGH THE DETECTION OF MICROORGANISMS

Foodborne pathogens have been known to be a responsible factor that affects the food safety most especially from the production stage, distribution, and during the ingestion of food. Food safety entails handling of food, their eventual preparation and storage in order to prevent related outbreaks and related illnesses. It has been recognized that biological hazards entails groups of pathogenic microorganisms that could altered the organoleptic features of but that could induce serious injuries mainly to the consumers. Also, foodborne diseases are mainly a result of pathogens that have been identified to cause long term influence on economic and social conditions leading to reduction in production (Plata, 2003). Specific examples of common foodborne pathogens includes *Salmonella spp,* and other disease-causing microorganisms (Torso et al., 2015).

Furthermore, numerous traditional as well as conventional techniques are utilized in the determination of foodborne pathogens. It has been discovered that the ELISA assay has been identified as one of the major techniques utilized for identification of pathogens present in animals, food, and tissues of humans (Wladir et al., 2015). This technique enhances proper detection of foodborne microorganisms because it takes considerable time and is not cost effective.

5.4.1 Salmonella spp.

Salmonella has been recognized as a major cause of salmonellosis, a typical example of foodborne diseases in humans and animals (Wang et al., 2011). It has been identified that annually about 155,000 deaths and 93.8 million human infections due to salmonellosis occurs globally (Hendriksen et al., 2011). Typical examples of the symptoms associated with these include fever and diarrhea (CDC, 2016). Hence, effective diagnoses of salmonella in a rapid and sensitive manner is imperative, most especially for food safety. It has been observed that salmonella spp thrives outsides its natural habitat, which includes food and water (White et al., 2002).

Several authors have utilized SPR-based experiments with antibodies for the isolation of *Salmonella spp* available in food items (Vaisocherova-Lisalova et al., 2016). Furthermore, it has been discovered that P-7 SPR-built optical fiber sensors could be utilized for the identification of salmonella (Romanov et al., 2011), while the application of novel DNA-based SPR biosensors could be utilized mainly for detection of genes of salmonella (Zhang et al., 2012). Also, network-based magnetic biosensors could be functional as a sensitive, quick and operative tool for easy identification of salmonella present on eggshells.

Garcia et al. (2012) stated the utilization of one-use DNA electrochemical platforms for rapid detection of *Salmonella spp*. Aptamers sensitive to nucleic acids have been identified as a single-stranded RNA or DNA molecules that possess unique binding affinities and structural forms for specific targets.

Ma et al. (2014) developed a biosensor that could detect the presence of salmonella. Moreover, electrochemical biosensing could be utilized for the detection of disease causing salmonella constructed on the utilization of a printed carbon electrode using carbon nanotubes, which led to the development of a disposable immunosensor. Park et al. (2013) utilized a phase-based magnetoelastic biosensors for the recognition of *Salmonella spp* available on a leaf of spinach.

5.4.2 Escherichia coli

E. coli O157:H7 has been identified as a dangerous isolate that could result in celiac diseases and has been identified to constitute a menace to human wellbeing. It has been identified to be responsible for the development of gastroenteritis and infant diarrhea. It has been observed to exist in food products, warm-blooded animals, and intestinal tracts of humans (Darnton et al., 2007). Some typical examples of infection as a result of *E. coli* includes kidney damage as a result of irritation in the small intestine and severe diarrhea with hemorrhage as well as bloodstream infections, respiratory illness, and urinary tract infections (CDC, 2014).

It has been observed that immunosensor design based on specific antibody-antigen could be utilized for identifying the existence of *Escherichia coli* (Leonard et al., 2003). Moreover, a fiber sensor has been discovered that does not require labeling, having a evanescent wave detection which could work based on changes in value of 280 nm, which is the light absorbance at target analyte could be utilized for the detection of *E. coli*.

It has also been discovered that electrochemical DNA biosensors could work based on magnetic beads which could decode the enzyme β-D-glucuronidase generated by *E. coli*. Furthermore, impedimetric sensing that works based on the covalently linked antibodies could work utilizing polyaniline film surface based on antibody-antigen binding techniques. These techniques have been shown to be inexpensive and patent-free which could detect the occurrence of *E. coli* in samples, and it is cost-effective (Chowdhury et al., 2012).

Also, three various types of electrodes have been designed using 3-aminopropyltriethoxysilane, carboxylic carbon nanotubes containing several walls and glutaraldehyde. These were designed to produce viable porous carbon paste electrodes that could easily detect *E. coli* (Lijian et al., 2012). The presence of *E. coli* O157:H7 in samples could be detected using an immunosensor developed based on the biofunctional magnetic beads that are said to portend that capacity to regulate the level of dose available in a nanoporous alumina membrane (Chan et al., 2013).

5.4.3 Listeria monocytogenes

This bacteria has been identified as a major factor responsible for listeriosis, which could lead to the development of infection that affects the placenta/fetus, blood, and central nervous tissue. In 2016, the Public Health Agency of Canada found that pregnant women and their developing children, together with people with low immune systems, possess the greater risk of illness that could be caused by Listeria (PHAC, 2016).

The application of impedimetric biosensors such as integrated biosensors utilize a combinatory effect of biological and impedance-recognition technologies with wide application for on-site detection. Liu and Zhang (2015) develop a covalently immobilized DNA probe that works based on hybridization reactions that were developed utilizing a paper-based microfluidic for rapid recognition of *L. monocytogenes*. It was discovered that the biosensor was very sensitive, and it is more reliable. Varshney et al. (2005) developed a new techniques called immunomagnetic separation that could remove the presence of pathogens mainly from food matrices.

5.4.4 Campylobacter spp.

It has been discovered that campylobacter possesses the ability to cause infection in humans through the consumption of highly infected meats. It has been discovered that bacteria have the potential to attack the digestive system of their host, causing infection of the digestive system and leading to the formation of campylobacteriosis. The symptoms of campylobacteriosis includes fever, diarrhea, abdominal pain and cramping. It has been stated that almost 2 million of cases as result of infection due to Campylobacter are normally recorded every year globally with 5–6 percent of circumstances linked to gastroenteritis (Medscape, 2015). Hence, there is a need to come up with an effective, rapid technology that could be utilized for the identification of Campylobacter spp.

It has been discovered that the application of Optical SPR biosensors could be utilized for the detection of Campylobacter (Wei et al., 2007). Aptamer investigation has gained research attention in recent times, most especially in the field of biosensors. Application of quantum dot fusion techniques and aptamer-magnetic bead DNA was done by means of aptamer sensors, in contrast to *Campylobacter spp* isolated surface proteins from $MgCl_2$ (Bruno et al., 2009).

Manzano et al. (2015) developed an organic deep-blue light-emitting diode that could be utilized for the development of a DNA biochip for determination of highly sensitive pathogens present on real fresh meat samples within a period of 24 hours (Manzano et al., 2015).

In the present day the position of biosensors cannot be overestimated. Food security agencies have zero tolerance for food contaminants, and biosensors have helped to check some problems associated with food security. Biosensors have helped to ensure that any of these pollutants are reduced from, or are entirely absent in food. Measures for testing food quality for global safety are currently available. This limited tolerance would be the essence of every novel biosensor in its plan and implementation to integrate detection inclusiveness and exclusiveness into systems.

5.5 CONCLUSION

In the present day the position of biosensors cannot be overestimated. Food security agencies have zero tolerance for food contaminants and it has been discovered that biosensors could help in checking some of these problems associated with food insecurity. This will go a long way in ensuring that the presence of these pollutants is reduced or entirely absent in food through provision of adequate measures for testing food quality in conjunction with global food-safety recommendations. This chapter gave comprehensive information on the utilization of biosensors for revealing isolates and several types of contaminants. Usage of various types of biosensors was also highlighted. There is an urgent need to invent a new biosensor that will be designed and implemented for proper integration in detection of different groups of contaminants, within the limited period of time, and without error.

REFERENCES

Adejumo, I.O., and Adetunji, C.O. 2018. Production and Evaluation of Biodegraded Feather Meal using Immobilised and Crude Enzyme from *Bacillus subtilis* on Broiler Chickens. *Brazilian Journal of Biological Sciences*, *5*(10): 405–416.

Adetunji, C.O., and Anani, O.A. 2021. Bioaugmentation: A Powerful Biotechnological Techniques for Sustainable Ecorestoration of Soil and Groundwater Contaminants. In: Panpatte, D.G., and Jhala, Y.K. (eds.), *Microbial Rejuvenation of Polluted Environment. Microorganisms for Sustainability*, vol. 25. Springer, Singapore. https://doi.org/10.1007/978-981-15-7447-4_15

Adetunji, C.O., Anani, O.A., Olaniyan, O.T., Inobeme, A., Olisaka, F.N., Uwadiae, E.O., and Obayagbona, O.N. 2021a. Recent Trends in Organic Farming. In: Soni, R., Suyal, D.C., Bhargava, P., and Goel, R. (eds.), *Microbiological Activity for Soil and Plant Health Management*. Springer, Singapore. https://doi.org/10.1007/978-981-16-2922-8_20

Adetunji, C.O., Egbuna, C., Oladosun, T.O., Akram, M., Michael, O., Olisaka, F.N., Ozolua, P., Adetunji, J.B., Enoyoze, G.E., and Olaniyan, O. 2021b. Efficacy of Phytochemicals of Medicinal Plants for the Treatment of Human Echinococcosis. Ch 8. In: *Neglected Tropical Diseases and Phytochemicals in Drug Discovery*. Wiley. DOI: 10.1002/9781119617143

Adetunji, C.O., Fawole, O.B., Afolayan, S.S., Olaleye, O.O., and Adetunji, J.B. 2012. *An Oral Presentation During 3rd NISFT Western Chapter Half Year Conference/General Meeting*, Ilorin, pp. 14–16.

Adetunji, C.O., Michael, O.S., Nwankwo, W., Ukhurebor, K.E., Anani, O.A., Oloke, J.K., Varma, A., Kadiri, O., Jain, A., and Adetunji, J.B. 2021c. Quinoa, The Next Biotech Plant: Food Security and Environmental and Health Hot Spots. In: Varma, A. (eds.), *Biology and Biotechnology of Quinoa*. Springer, Singapore. https://doi.org/10.1007/978-981-16-3832-9_19

Adetunji, C.O., Olaniyan, O.T., Adeyomoye, O., Dare, A., Adeniyi, M.J., Alex, E., Rebezov, M., Isabekova, O., and Ali Shariati, M. 2022a. Smart Sensing for COVID-19 Pandemic. In: Pani, S.K., Dash, S., dos Santos, W.P., Chan Bukhari, S.A., Flammini, F. (eds), *Assessing COVID-19 and Other Pandemics and Epidemics using Computational Modelling and Data Analysis*. Springer, Cham. https://doi.org/10.1007/978-3-030-79753-9_9

Adetunji, C.O., Olugbemi, O.T., Akram, M., Laila, U., Samuel, M.O., Oshinjo, A.M., Adetunji, J.B., Okotie, G.E., (Diuto) Esiobu, N., Oyedara, O.O., and Adeyemi, F.M. 2022b. Application of Computational and Bioinformatics Techniques in Drug Repurposing for Effective Development of Potential Drug Candidate for the Management of COVID-19. In: *Medical Biotechnology, Biopharmaceutics, Forensic Science and Bioinformatics*. CRC Press, p. 14. eBook ISBN 9781003178903. DOI: 10.1201/9781003178903-15

Adetunji, C.O., Mitembo, W.P., Egbuna, C., and Narasimha Rao, G.M. 2020. Silico Modeling as a Tool to Predict and Characterize Plant Toxicity. In: Mtewa, A.G., Egbuna, C., Narasimha Rao, G.M. (eds.), *Poisonous Plants and Phytochemicals in Drug Discovery*. Wiley Online Library. https://doi.org/10.1002/9781119650034.ch14.

Adetunji, C.O., Nwankwo, W., Olayinka, A.S., Olugbemi, O.T., Akram, M., Laila, W., Samuel, M.O., Oshinjo, A.M., Adetunji, J.B., Okotie, G.E., and (Diuto) Esiobu, N. 2022c. Computational Intelligence Techniques for Combating COVID-19. In: *Medical Biotechnology, Biopharmaceutics, Forensic Science and Bioinformatics*. CRC Press, p. 12. eBook ISBN 9781003178903. DOI: 10.1201/9781003178903-16

Adetunji, C.O., Nwankwo, W., Olayinka, A.S., Olugbemi, O.T., Akram, M., Laila, U., Olugbenga, M.S., Oshinjo, A.M., Adetunji, J.B., Okotie, G.E., and (Diuto) Esiobu, N. 2022d. Machine Learning and Behaviour Modification for COVID-19. In: *Medical Biotechnology, Biopharmaceutics, Forensic Science and Bioinformatics*. CRC Press, p. 17. eBook ISBN 9781003178903. DOI: 10.1201/9781003178903-17.

Adetunji, C.O., Olaniyan, O.T., Adeyomoye, O., Dare, A., Adeniyi, M.J., Alex, E., Rebezov, M., Garipova, L., and Ali Shariati, M. 2022e. eHealth, mHealth, and Telemedicine for COVID-19 Pandemic. In: Pani, S.K., Dash, S., dos Santos, W.P., Chan Bukhari, S.A., and Flammini, F. (eds.), *Assessing COVID-19 and Other Pandemics and Epidemics using Computational Modelling and Data Analysis*. Springer, Cham. https://doi.org/10.1007/978-3-030-79753-9_10

Adetunji, C.O., Olaniyan, O.T., Adeyomoye, O., Dare, A., Adeniyi, M.J., Alex, E., Rebezov, M., Petukhova, E., and Ali Shariati, M. 2022f. Machine Learning Approaches for COVID-19 Pandemic. In: Pani, S.K., Dash, S., dos Santos, W.P., Chan Bukhari, S.A., and Flammini, F. (eds.). *Assessing COVID-19 and Other Pandemics and Epidemics using Computational Modelling and Data Analysis*. Springer, Cham. https://doi.org/10.1007/978-3-030-79753-9_8

Adetunji, C.O., Olaniyan, O.T., Adeyomoye, O., Dare, A., Adeniyi, M.J., Alex, E., Rebezov, M., Petukhova, E., and Ali Shariati, M. 2022g. Internet of Health Things (IoHT) for COVID-19. In: Pani, S.K., Dash, S., dos Santos, W.P., Chan Bukhari, S.A., and Flammini, F. (eds.), *Assessing COVID-19 and Other Pandemics and Epidemics using Computational Modelling and Data Analysis*. Springer, Cham. https://doi.org/10.1007/978-3-030-79753-9_5

Adetunji, C.O., Olaniyan, O.T., Adeyomoye, O., Dare, A., Adeniyi, M.J., Alex, E., Rebezov, M., Koriagina, N., and Ali Shariati, M. 2022h. Diverse Techniques Applied for Effective Diagnosis of COVID-19. In: Pani S.K., Dash S., dos Santos W.P., Chan Bukhari S.A., and Flammini F. (eds.), *Assessing COVID-19 and Other Pandemics and Epidemics using Computational Modelling and Data Analysis*. Springer, Cham. https://doi.org/10.1007/978-3-030-79753-9_3

Adetunji, C.O. and Oyeyemi, O.T. 2022i. Antiprotozoal Activity of Some Medicinal Plants against Entamoeba Histolytica, the Causative Agent of Amoebiasis. In: *Medical Biotechnology, Biopharmaceutics, Forensic Science and Bioinformatics*. CRC Press, p. 12. eBook ISBN 9781003178903. www.taylorfrancis.com/chapters/edit/10.1201/9781003178903-20/antiprotozoal-activity-medicinal-plants-entamoeba-histolytica-causative-agent-amoebiasis-charles-oluwaseun-adetunji-oyetunde-oyeyemi.

Adetunji, C.O., and Ukhurebor, K.E. 2021. Recent Trends in Utilization of Biotechnological Tools for Environmental Sustainability. In: Adetunji, C.O., Panpatte, D.G., and Jhala, Y.K. (eds.), *Microbial Rejuvenation of Polluted Environment. Microorganisms for Sustainability*, vol. 27. Springer, Singapore. https://doi.org/10.1007/978-981-15-7459-7_11

Adley, C. 2014. Past, Present and Future of Sensors in Food Production. *Foods*, 2014, 3: 491–510; doi:10.3390/foods3030491, www.mdpi.com/journal/foods, ISSN 2304-8158.

Akram, M., Adetunji, C.O., Egbuna, C., Jabeen, S., Olaniyan, O., Ezeofor, N.J., Anani, O.A., Laila, U., Găman, M.-A., Patrick-Iwuanyanwu, K., Ifemeje, J.C., Chikwendu, C.J., Michael, O.C., and Rudrapal, M. 2021a. Dengue Fever, Chapter 17. In: *Neglected Tropical Diseases and Phytochemicals in Drug Discovery*. Wiley. DOI: 10.1002/9781119617143

Akram, M., Mohiuddin, E., Adetunji, C.O., Oladosun, T.O., Ozolua, P., Olisaka, F.N., Egbuna, C., Michael, O., Adetunji, J.B., Hameed, L., Awuchi, C.G., Patrick-Iwuanyanwu, K., and Olaniyan, O. 2021b. Prospects of Phytochemicals for the Treatment of Helminthiasis. In: *Neglected Tropical Diseases and Phytochemicals in Drug Discovery*. Chapter 7: Wiley. DOI: 10.1002/9781119617143

Alocilja, E.C. and Radke, S.M. 2003. Market Analysis of Biosensors for Food Safety. *Biosensors and Bioelectronics* 18(5–6): 841–846.

Ampuero, S. and Bosset, J. 2003. The Electronic Nose Applied to Dairy Products: A Review. *Sensors Actuators B: Chemistry* 94(1): 1–12.

Arowora, K.A., Abiodun, A.A., Adetunji, C.O., Sanu, F.T., Afolayan, S.S., and Ogundele, B.A. (2012). Levels of Aflatoxins in Some Agricultural Commodities Sold at Baboko Market in Ilorin, Nigeria. *Global Journal of Science Frontier Research* 12(10): 31–33.

Bahadır, E.B., and Sezgintürk, M.K. 2015. Applications of Electrochemical Immunosensors for Early Clinical Diagnostics. *Talanta* 132:162–174. doi: 10.1016/j.talanta.2014.08.063.

Bruno J.G., Phillips, T., Carrillo, M.P., and Crowell, R. 2009. Plastic-adherent DNA Aptamer-magnetic Bead and Quantum Dot Sandwich Assay for Campylobacter Detection. *Journal of Fluorescence* 19(3):427–435.

CDC. 2014. Shiga Toxin-Producing *E. coli* & Food Safety. Centers for Disease Control and Prevention. Available at: www.cdc.gov/Features/EcoliInfection/index.html

CDC. 2016. 2015 Food Safety Report. Centers for Disease Control and Prevention. Available at: www.cdc.gov/foodnet/index.html

Chai, Y., Li, S., Horikawa, S., Park, M.-K.,Vodyanoy, V., and Bryan, A. 2012. Rapid and Sensitive Detection of *Salmonella typhimurium* on Eggshells by Using Wireless Biosensors. *Journal of Food Protection* 75(4):631–636.

Chan, K.Y., Ye, W.W., Zhang, Y., Xiao, L.D., Leung, P.H.M., Li, Y., and Yang, M. 2013. Ultrasensitive Detection of *E. coli* O157:H7 with Biofunctional Magnetic Bead Concentration via Nanoporous Membrane Based Electrochemical Immunosensor. *Biosensors and Bioelectronics* 41:532–537.

Chowdhury, A.D., De, A., Chaudhuri, C.R., Bandyopadhyay, K., and Sen, P. 2012. Label Free Polyaniline Based Impedimetric Biosensor for Detection of *E. coli* O157: *H7 Bacteria. Sensors and Actuators B: Chemical* 171–172:916–923.

Darnton, N., Turner, L., Rojevsky, S., and Berg, H. 2007. On Torque and Tumbling in Swimming *Escherichia coli. Journal of Bacteriology* 189(5):1756–1764.

Dauda, W.P., Abraham, P., Glen, E., Adetunji, C.O., Ghazanfar, S., Ali, S., Al-Zahrani, M., Azameti, M.K., Alao, S.E.L., Zarafi, A.B., Abraham, M.P., and Musa, H. 2022a. Robust Profiling of Cytochrome P450s (P450ome) in Notable *Aspergillus* spp. *Life* 12(3): 451. https://doi.org/10.3390/life12030451

Dauda, W.P., Morumda, D., Abraham, P., Adetunji, C.O., Ghazanfar, S., Glen, E., Abraham, S.E., Peter, G.W., Ogra, I.O., Ifeanyi, U.J., Musa, H., Azameti, M.K., Paray, B.A., and Gulnaz, A. 2022b. Genome-wide Analysis of Cytochrome P450s of Alternaria Species: Evolutionary Origin, Family Expansion and Putative Functions. *Journal of Fungi* 8(4): 324. https://doi.org/10.3390/jof8040324

Esser, B., Schnorr, J.M., and Swager, T.M. 2012. Selective Detection of Ethylene Gas using Carbon Nanotube-based Devices: Utility in Determination of Fruit Ripeness. *Angewandte Chemie International Edition* 51: 5752–5756.

Gallo, M., and Ferranti, P. 2016. The Evolution of Analytical Chemistry Methods in Foodomics *Journal of Chromatography A* 1428: 3–15.

Garcia, T., Revenga-Parra, M., Anorga, L., Arana, S., Pariente, F., and Lorenzo, E. 2012. Disposable DNA Biosensor Based on Thin-film Gold Electrodes for Selective Salmonella Detection. *Sensors and Actuators B: Chemical* 161(1): 1030–1037.

Handa, P. and Singh, B. 2018. Importance of Electronics in Food Industry. *IOSR Journal of Electronics and Communication Engineering* 13(2), Ver. II (March–April 2018), pp. 15–22. www.iosrjournals.org, Doi: 10.9790/2834-1302021522

Hendriksen, R.S., Vieira, A.R., Karlsmose, S., Wong, D.M.A.L.F., Jensen, A.B., Wegener, H.C., and Aarestrup, F.M. 2011. Global monitoring of Salmonella Serovar distribution from the World Health Organization Global Foodborne Infections Network Country Data Bank: Results of Quality Assured Laboratories from 2001 to 2007. *Foodborne Pathogens and Disease* 8(8): 887–900.

Hoffmann, S., Maculloch, B., and Batz, M. 2015. *Economic Burden of Major Foodborne Illnesses Acquired in the United States. USDA-140.* Washington, DC.

Jiang, H. Jiang, D., Zhu, P., Pi, F., Ji, J., Sun, C., Sun, J., and Sun, X. 2016. A Novel Mast Cell Co-culture Microfluidic Chip for the Electrochemical Evaluation of Food Allergen. *Biosensors and Bioelectronics* 83:126–133.

Jones, P., Clarke-Hill, C., Hillier, D., and Comfort, D. 2005. The Benefits, Challenges And Impacts Of Radio Frequency Identification Technology (RFID) for Retailers in the UK. *Marketing Intelligence & Planning* 23:395–402.

Joyner, J.R. and Kumar, D.V. 2015. Nanosensors and Their Applications in Food Analysis: A Review 1(4): 80–90

Leonard, P., Hearty, S., Brennan, J., Dunne, L., Quinn, J., Chakraborty, T., and O'Kennedy, R. 2003. Advances in Biosensors for Detection of Pathogens in Food and Water. *Enzyme Microbial Technology* 32(1): 3–13.

Lijian, X., Jingjing, D., Yan, D., and Nongyue, H. 2012. Electrochemical Detection of *E. coli* O157:H7 Using Porous Pseudo-carbon Paste Electrode Modified with Carboxylic Multi-walled Carbon Nanotubes, Glutaraldehyde and 3-Aminopropyltriethoxysilane. *Journal of Biomedical Nanotechnology* 8(6): 1006–1011.

Liu, F. and C. Zhang. 2015. A Novel Paper-based Microfluidic Enhanced Chemiluminescence Biosensor for Facile, Reliable and Highly-Sensitive Gene Detection of Listeria Monocytogenes. *Sensors and Actuators B: Chemical* 209: 399–406.

Lorenzo, J.M., Munekata, P.E., Dominguez, R., Pateiro, M., Saraiva, J.A., and Franco, D. 2017. Main Groups of Microorganisms of Relevance for Food Safety and Stability: General Aspects and Overall Description. *Innovative Technologies for Food Preservation* 2018: 53–107. doi: 10.1016/B978-0-12-811031-7.00003-0.

Ma, X., Jiang, Y., Jia, F., Yu, Y., Chen, J. and Wang, Z. 2014. An Aptamer-based Electrochemical Biosensor for the Detection of Salmonella. *Journal of Microbiological Methods* 98: 94–98.

Maduraiveeran, G. and Jin, W. 2017. Nanomaterials Based Electrochemical Sensor and Biosensor Platforms for Environmental Applications. *Trends in Environmental Analytical Chemistry* 13:10–23. https://doi.org/10.1016/j.teac.2017.02.001

Manzano, M., Cecchini, F., Fontanot, M., Iacumin, L., Comi, G., and Melpignano, P. 2015. OLED-based DNA Biochip for Campylobacter spp. Detection in Poultry Meat Samples. *Biosensors and Bioelectronics* 66: 271–276.

MedScape. 2015. Campylobacter Infections. Available at: emedicine.medscape.com/article/213720-overview#a6

Mustafa, F. and Andreescu, S. 2018. Chemical and Biological Sensors for Food-Quality Monitoring and Smart Packaging. *Foods* 7(10):168.

Neethirajan, S., Ragavan, V., Weng, X., and Chand, R. (2018). Biosensors for Sustainable Food Engineering: Challenges and Perspectives. *Biosensors 8*(1): 23.

Nwankwo, W., Adetunji, C.O., Ukhurebor, K.E., Panpatte, D.G., Makinde, A.S., and Hefft, D.I. (2021). Recent advances in application of microbial enzymes for biodegradation of waste and hazardous waste material. In: Adetunji, C.O., Panpatte, D.G., and Jhala, Y.K. (eds.), *Microbial Rejuvenation of Polluted Environment. Microorganisms for Sustainability*, vol. 27. Springer, Singapore. https://doi.org/10.1007/978-981-15-7459-7_3

Okeke, N.E., Adetunji, C.O., Nwankwo, W., Ukhurebor, K.E., Makinde, A.S., and Panpatte, D.G. (2021). A Critical Review of Microbial Transport in Effluent Waste and Sewage Sludge Treatment. In: Adetunji, C.O., Panpatte, D.G., Jhala, Y.K. (eds.), *Microbial Rejuvenation of Polluted Environment. Microorganisms for Sustainability*, vol. 27. Springer, Singapore. https://doi.org/10.1007/978-981-15-7459-7_10

Olaniyan, O.T., Adetunji, C.O., Adeniyi, M.J., and Hefft, D.I. (2022). Computational Intelligence in IoT Healthcare. In: *Deep Learning, Machine Learning and IoT in Biomedical and Health Informatics.* CRC Press, p. 13. eBook ISBN 9780367548445. DOI: 10.1201/9780367548445-19

Oluwaseun, A.C., Phazang, P., and Sarin, N.B. 2017. Significance of Rhamnolipids as a Biological Control Agent in the Management of Crops/Plant Pathogens. *Current Trends in Biomedical Engineering & Biosciences* 10(3): 54–55. Juniper Publishers Inc.

Omanović-Mikličanina, E. and Maksimović, M. 2016. Nanosensors Applications in Agriculture and Food Industry. *Bulletin of the Chemists and Technologists of Bosnia and Herzegovina* 47:59–70.

Oyedara, O.O., Adeyemi, F.M., Adetunji, C.O., Elufisan, T.O. (2022). Repositioning Antiviral Drugs as a Rapid and Cost-effective Approach to Discover Treatment against SARS-CoV-2 Infection. In: *Medical Biotechnology, Biopharmaceutics, Forensic Science and Bioinformatics*. CRC Press, p. 12. eBook ISBN 9781003178903. DOI: 10.1201/9781003178903-10.

Park, M.K., Park, J. W., Wikle, H.C., III, and Chin, B.A. 2013. Evaluation of Phage-based Magnetoelastic Biosensors for Direct Detection of *Salmonella typhimurium* on Spinach Leaves. *Sensors and Actuators B: Chemical* 176:1134–1140.

PHAC. 2016. Public Health Notice Update – Outbreak of Listeria Infections Linked to Packaged Salad Products Produced at the Dole Processing Facility in Springfield, Ohio: Public Health Agency of Canada. Available at: www.phac-aspc.gc.ca/phn-asp/2016/listeria-eng.php.

Pividori, I. and Alegret, S. 2010. Electrochemical Biosensors for Food Safety. *Contributions to Science* 6(2): 173–191 2010. DOI: 10.2436/20.7010.01.95 ISSN: 1575–6343www.cat-science.cat

Plata, G.V.D. 2003. La Contaminacion de los Alimentos, unProblemapor Resolver. *Salud UIS* 35(1): 48–57 (In Spanish, with English abstract).

Priyanka, H., and Bhupinder, S. 2016. Electronic Nose and Their Application in Food Industries. *Food Science Research Journal* 7(2): 314–318.

Romanov, V., Galelyuka, I., Glushkov, V., Starodub, N., and Son'ko, R. 2011. P7 – Optical Immune Biosensor Based on SPR for the Detection of *Salmonella typhimurium*. In: *Proceedings OPTO 2011*, pp.139–144, Nurnberg, AMA Conferences.

Saleem, M. 2013. Biosensors a Promising Future in Measurements. *IOP Conference Series: Materials Science and Engineering* 51:012012.

Sangeetha, J., Hospet, R., Thangadurai, D., Adetunji, C.O., Islam, S., Pujari, N., and Al-Tawaha, A.R.M.S. 2021. Nanopesticides, Nanoherbicides, and Nanofertilizers: The Greener Aspects of Agrochemical Synthesis Using Nanotools and Nanoprocesses Toward Sustainable Agriculture. In: Kharissova, O.V., Torres-Martínez, L.M., and Kharisov, B.I. (eds.), *Handbook of Nanomaterials and Nanocomposites for Energy and Environmental Applications*. Springer, Cham. https://doi.org/10.1007/978-3-030-36268-3_44

Soh, N., Tokuda, T., Watanabe, T., Mishima, K., Imato, T., Masadome, T., Asano, Y., Okutani, S., Niwa, O., and Brown, S. 2003. A Surface Plasmon Resonance Immunosensor for Detecting a Dioxin Precursor using a Gold Binding Polypeptide. *Talanta* 60(4): 733–745. doi: 10.1016/S0039-9140(03)00139-5. PMID: 18969098.

Takhistov, P. 2005. Biosensor Technology for Food Processing, Safety and Packaging. 45(2): 231–240.

Thakur, M.S. and Ragavan, K.V. 2013. Biosensors in Food Processing. *Journal of Food Science and Technology* 50(4): 625–641.

Torso, L.M., Voorhees, R.E., Forest, S.A., Gordon, A.Z., Silvestri, S.A., Kissler, B., Schlackman, J., Sandt, C.H., Toma, P., Bachert, J., Mertz, K.J., and Harrison, L.H. 2015. *Escherichia coli* O157:H7 Outbreak Associated with Restaurant Beef Grinding. *Journal of Food Protection* 78(7):1272–1279.

Ukhurebor, K.E., Adetunji, C.O., Bobadoye, A.O., Aigbe, U.O., Onyancha, R.B., Siloko, I.U., Emegha, J.O., Okocha, G.O., and Abiodun, I.C. (2021). Bionanomaterials for Biosensor Technology. *Bionanomaterials. Fundamentals and Biomedical Applications*, 5–22.

Ukhurebor, K.E., Mishra, P., Mishra, R.R., and Adetunji, C.O. (2020). Nexus between climate change and food innovation technology: Recent advances. In: Mishra, P., Mishra, R.R., and Adetunji, C.O. (eds.), *Innovations in Food Technology*. Springer, Singapore. https://doi.org/10.1007/978-981-15-6121-4_20

Ukhurebor, K.E., Nwankwo, W., Adetunji, C.O., and Makinde, A.S. (2021). Artificial Intelligence and Internet of Things in Instrumentation and Control in Waste Biodegradation Plants: Recent Developments. In: Adetunji, C.O., Panpatte, D.G., and Jhala, Y.K. (eds.), *Microbial Rejuvenation of Polluted Environment. Microorganisms for Sustainability*, vol. 27. Springer, Singapore. https://doi.org/10.1007/978-981-15-7459-7_12

Vaisocherova-Lisalova, H., Višova, I., Ermini, M.L., Špringer, T., Song, X.C., Mrazek, J., Lamačova, J., Lynn, N.S., Šedivak, P., and Homol, J. 2016. Low-fouling Surface Plasmon Resonance Biosensor for Multi-step Detection of Foodborne Bacterial Pathogens in Complex Food Samples. *Biosensors and Bioelectronics* 80: 84–90.

Varshney, M., Yang, L., Su, X.L., and Li, Y. 2005. Magnetic Nanoparticle-Antibody Conjugates for the Separation of *Escherichia coli* 0157:H7 in Ground Beef. *Journal of Food Protection* 68: 1804–1811.

Wang, Q., Fang, J., Cao, D., Li, H., Su, K., Hu, N., and Wang, P. 2015. An Improved Functional Assay for Rapid Detection of Marine Toxins, Saxitoxin and Brevetoxin using a Portable Cardiomyocyte-Based Potential Biosensor. *Biosensors and Bioelectronics* 72: 10–17.

Wang, Y., Zhang, S., Du, D., Shao, Y., Li, Z., Wang, J., Engelhard, M.H., Li, J., and Lin, Y. 2011. Self-assembly of Acetylcholinesterase on a Gold Nanoparticles-Graphene Nanosheet Hybrid for Organophosphate Pesticide Detection Using Polyelectrolyte as a Linker. *Journal of Materials Chemistry* 21(14): 5319–5325.

Wei, D., Oyarzabal, O., Huang, T., Balasubramanian, S., Sista, S., and Simonian, A. 2007. Development of a Surface Plasmon Resonance Biosensor for the Identification of *Campylobacter jejuni*. *Journal of Microbiological Methods* 69(1): 78–85.

White, D.G., Zhao, S., Simjee, S., Wagner, D.D., and McDermott, P.F. 2002. Antimicrobial Resistance of Foodborne Pathogens. *Microbes and Infection* 4(4): 405–412.

WHO. 2015. *World Health Day 2015: Food Safety*. India.

Wladir, B.V., Edward, G.D., Doores, S., and Cutter, C.N. 2015. Commercially Available Rapid Methods for Detection of Selected Foodborne Pathogens. *Food, Science and Nutrition* 56(9): 1519–1531.

Xing, G., Sang, S., Guo, J., Jian, A., Duan, Q., Ji, J., Zhang, Q., and Zhang, W. 2017. A Magnetoelastic Biosensor Based on E2 Glycoprotein for Wireless Detection of Classical Swine Fever Virus E2 Antibody. *Science Report* 7: 15626. https://doi.org/10.1038/s41598-017-15908-2

Zhang, L., Zhang, A., Du, D., and Lin, Y. 2012. Biosensor Based on Prussian Blue Nanocubes/Reduced Grapheme Oxidenanocomposite for Detection of Organophosphorus Pesticides. *Nanoscale* 4(15): 4674–4679.

6 Packaging of Animal-Based Food Products

The Need for Intelligent Freshness Sensors

Ogundolie Frank Abimbola, Charles Oluwaseun Adetunji, Babatunde Joseph Dare, John Tsado Mathew, Abel Inobeme, Shakira Ghazanfar, Olugbemi T. Olaniyan, Modupe Doris Ajiboye, and Wadazani Dauda

CONTENTS

DOI: 10.1201/9781003207955-6

6.1 INTRODUCTION

In the past few years, there has been a surge in awareness in the usage of innovative packaging technologies, both for meat products and meat. The use of substances in packaging methods to preserve or increase the shelf-life, organoleptic properties, nutritive values and quality of meat products is known as novel packaging (Risch and Sara, 2009).

Oxygen scavengers, emitters and carbon dioxide scavengers' moisture-regulating substances, and food packaging materials techniques are among the dynamic packaging technologies explored (Smith et al., 2013). Efficient packaging devices monitor the state of packaged foods to provide information about their viability throughout storage as well as transport (Arvanitoyannis, 2005). For prospective usage in meat products and meat, indicators (particularly freshness, time-temperature indicators (TTI) and integrity), sensor technologies, and radio frequency identification are explored. The food sector must recognize the benefits of intelligent and active packaging technologies, as well as produce economically feasible packaging methods and enhance consumer acceptability, for such innovations to be commercially successful (Kerry et al., 2006; Yildirim et al. 2017).

Consumers are increasingly demanding fresh, high-quality, and healthy food. As a result, innovative packaging mechanisms, such as intelligent packaging, have emerged. Good packaging embedded indicators or sensors offer more information about the time-temperature and integrity history of the packaged foods, which is useful for ensuring the quality and safety as well as freshness of packaged meals (Adetunji et al., 2017; Olayemi and Adetunji, 2013; Adetunji et al., 2013; Adetunji et al., 2021a,b,c; Adetunji et al., 2022 a,b,c,d,e,f,g). Although food packaging's primary functions seem to be to protect and preserve food, packaging also provides information about the food quality or freshness, tamper indications, safety or traceability via indicators or sensors (Robertson, 2013).

As a result, intelligent or smart packaging enables manufacturers to supply fresh, safe items and high-quality to customers. A sensor, which could be characterized as a tool or instrument that determines, detects, or quantifies matter or energy and provides a signal or response for the assessment or quantification of physical or chemical property, is utilized in sustainable packaging. Sensors have two fundamental components: a transducer and a receptor, and combined input signals, although the indicator is a related phrase that is also utilized in packaging technologies. An indicator is a substance or material that uses a fundamental variation, including mass or energy/color, to confirm the absence, existence, or intensity of another substance as little more than an indicator, or even the amount of reactivity among two or more substances. Although the concepts are frequently interchanged, indicator usually refers to a spectrophotometric sensor that could either be a biosensor a chemical (Kuswandi, 2017).

Aseptic packaging techniques are rejected because they do not give real-time information to the manufacturers and consumers concerning the quality of packaged food products along the supply chain. To prevent dangers, such as the essence of a modern technology to control food rotting through the field to the table, has evolved as have foodborne diseases (Adeyemi et al. 2017; Ogundolie et al. 2022). Furthermore, food standards surveillance systems reveal the primary causes of food waste in the production process. To improve the food quality of the product and match consumer expectations, smart packaging is used to convey an overview of the history of food storage and handling. Meat is among the most perishable commodities, and consuming spoiled meat products can result in serious infections. In the meat sector, an array of indicators and sensors are developed to alert consumers of the status of packaged meat products, including their level of deterioration.

In the past few years, biosensor technology and the chemical sensor have advanced fast. Optical, electrical, chemical and thermal signal domains are the most common types of sensors that could be used in meat packaging technologies. Sensors could be used to determine the value of a principal measurement parameter or, mostly in the case of the marker approach, to determine the value from another chemical, biological or physical variable. Appropriate readings are required as predictors of the meat quality of the product in the instance of atmospheric sensing applications. Although the gap among conceptual and economically viable sensing devices has reduced, and while useful applications of sensors mostly in the meat industry remain stable, major effective measures more toward general utilization have been achieved.

The purpose of preservative packaging for processed meat is to preserve an acceptable odor, appearance, and flavor, as well as to stabilize content and prevent microbiological decomposition. For food products, such as cooked as well as fresh meats and processed meats, several packaging solutions and techniques are presently accessible. Variable factors observed in packaged foods such as in the headspace include product ratio in the packaged foods, the oxygen concentration, storage temperature, pH change, moisture content, and product composition (Kerry and Butler, 2008; Thakur and Ragavan, 2013; Mehrotra, 2016), and influences the color stability in processed foods, which also eventually affects general consumer acceptance of this food.

Meat is a fragile food that spoils quickly if kept out in the open and uncooked at room temperature. Meat may only be saved for future use if it is processed, packaged, and stored properly. Rising urbanization, as well as lifestyle changes, have increased the need for not only ready-to-eat meat but also easy meat products. However, meat is an ideal setting for bacteria metabolism given its biological features and chemical composition. The early microbiome is diverse, with psychotropic and mesophilic bacteria that can cause illnesses in people along with animals, as well as meat deterioration. To set aside or limit good manipulation procedures, adequate conservation measures and microbial growth are required. The present integrated conservation method relies on refrigeration and several kinds of packages to achieve a maximum shelf span (Mathew and Jaganathan, 2017).

The development of active food packaging systems helps in addressing several concerns in the meat packaging industry. This new packaging system ensures the safety of the food, ensures that packaged food is of high quality and monitored along the production and supply chains (Bi, 2012). This further increases the acceptance and confidence of individual consumers towards healthy and good packaged products. The use of intelligent and active packaging technologies provides a variety of options aimed at improving food safety, quality and shelf-life of packaged products.

Also novel, there are groundbreaking operating procedures aimed at improving pH, thermal stability, mechanical strength and barrier performance of inactive food packaged. These procedures are improving consumers' faith in animal-based packaged food today. The utilization of biodegradable or edible materials, nanomaterials along plant extracts in the production of green and sustainable packaging seems to have the promise to lessen the ecological implications of packaged foods. Intelligent, active, but also green, packaging innovations could all work together to create a multifunctional food supply chain with little or no negative individual constituent, and this could be considered the purpose of packaging material for future technology.

The above review examines the fundamentals of food packaging as well as current advancements in various kinds of food packaging systems (Han et al., 2018). The purpose of this chapter is to provide issues and potential methodologies and also available technologies for monitoring beef quality throughout transportation and storage. In addition, emerging applications are compared in terms of benefits and drawbacks in meatpacking uses (Mohebi and Marquez, 2015).

6.2 ACTIVE PACKAGING SYSTEMS IN ANIMAL-BASED FOOD PACKAGING

The impact of oxygen, oxygen radical, and metallic ions on preserved foods include but is not limited to the following rancidity of unsaturated fats, promoting the growth of microbes (bacteria and fungi), darkening of fresh pigments, loss of flavor, and phenolic browning of fruit/vegetables. The alteration in contents and compositions observed in preserved food such as loss of flavor in beer, vitamin C (ascorbic acid), and degradation and loss of aroma increases food spoilage with measurable economic damages. However, an active packaging system using technological-based approaches in packaging food would provide solutions to the microbial growth and rancidity formation in food products that will promote food security.

6.3 OXYGEN SCAVENGERS AS AN INTELLIGENT SENSOR IN MAINTAINING ANIMAL-BASED FOOD PACKAGING INTEGRITY

Oxygen scavengers have been recognized to actively inhibit food microbes that initiate degradation of food contents and compositions, thereby reducing the quality and the integrity of the packaged food materials. By this singular act, oxygen scavengers play an important role in controlling the dissolved oxygen radicals, and maintaining the color, texture and composition of different food products. Free radicals distort the composition and the contents of the DNA, RNA, proteins, fats and oil, and carbohydrates molecules. Alterations in these micro and macromolecules underline the distortions in food components and the consequent spoilages observed. This is because the oxygen radicals are highly unstable and reactive, a typical characteristic of the oxidants or reductants, such as hydroxyl radical, superoxide anion radical, hydrogen peroxide, oxygen radicals, hypochlorite, nitric oxide radical, and peroxynitrite radicals.

These oxygen scavengers alter the chain of reaction in the degradation mechanism by donating an electron to the free radicals present in the food-packaging systems as well as causing removal of reactive oxygen species/reactive nitrogen species initiators. Another mode of action of oxygen scavengers includes metal ion chelation, co-antioxidants, or gene expression regulation. These substances act as radical scavengers, hydrogen donors, electron donors, peroxide decomposers, singlet oxygen quenchers, enzyme inhibitors, synergists, and metal-chelating agents.

Oxygen scavengers reduce hydroperoxides and hydrogen peroxide to alcohols and water, with no further production of free radicals, this gives the primary understanding to the mechanism of action in preserving packaged food integrity in terms of quality and maintaining the state of the freshness of the packaged food system

Metal toxicity obtained from exposure to heavy metals and other sources during the food-packaging system underline the physiological and pathological processes of food damage and in many diseases resulting in metal poisoning. However, when present in the packaging materials, chelating agents bind to ions produced; some of these ions are toxic at either low or high concentrations. The binding/chelation of these agents results in the formation of non-toxic EDTA-ion complexes which are easy to remove from the food-packaging system, examples of such chemical substances include; Ethylenediamine-tetraacetic acid ($EDTA^{4-}$), a hexadentate ligand as a lead chelating agent, dimercaprol (British Anti-Lewisite or BAL), as a chelating agent for arsenic and mercury poisoning, Desferrioxamine (DFOA), an iron chelator.

Ferrous iron-based oxygen scavengers are common in the preservation of packaged foods. Oxygen absorbers are packaged in paper and polyethene, safe to use, not edible, and non-toxic. There is no production of harmful gases. After use, oxygen absorber package systems form a red rusty color, harder to feel from the exterior.

6.4 CARBON DIOXIDE SCAVENGERS AND EMITTERS AS AN INTELLIGENT SENSOR IN MAINTAINING ANIMAL-BASED FOOD PACKAGING INTEGRITY

In the modified atmosphere packaging (MAP) system (Lee, 2016), also a type of smart packaging system, the shelf-life, freshness and quality of packaged foods are preserved by the introduction of carbon dioxide, which is introduced by flushing. This gas at a high concentration inhibits the growth of microbes, thereby preventing spoilage. The quality of packaged food is usually maintained when the carbon dioxide level in the packaged product is low; this provides the best preservation concentration but is largely affected adversely when there is an excess accumulation of CO_2. This affects the integrity and nutritive value of the packaged food product. In some cases CO_2 emitters are required to mitigate relatively high carbon dioxide in food packages. Also, this gas has been reported to be effective against microbes, either when present alone or in the presence of nitrogen, when it has a better effect.

The concentration of carbon dioxides is important in ensuring adequate preservation as excess CO_2 needs to be removed from the system (Puligundla et al., 2012). In the industry, CO_2 absorbers are included inside the packaging material to help remove excess CO_2 present in the headspace of the packaging material. CO_2 absorbers can either be physical or chemical (Lee, 2016). The common CO_2 absorbers that have presently been used in various industries include magnesium hydroxide ($Mg(OH)_2$), calcium hydroxide ($Ca(OH)_2$), amino-acid salt solutions sodium chloride (NaCl), sodium carbonate (Na_2CO_3) and calcium oxide, CaO (Lee, 2016).

In the packaging of products, the use of crystalline alumino-silicates and activated carbon, silica gels which are generally referred to as physical adsorbents are usually in powdered or granules and are designed within the packaging films or materials (Saha, Jribi, Koyama, and El-Sharkawy, 2011; Hauchhum and Mahanta, 2014). These physical adsorbents help in absorbing excess carbon dioxide present in the packaged material.

6.5 BENEFITS OF ACTIVE PACKAGING TO PRESERVED FOODS

Packaging foods in the industry is aimed at limiting contamination, which can arise as a result of environmental factors. Food spoilage has played an important role over the years in economic losses to various food-based industries, either spoilage caused as a result of transportation, chemical contamination, environmental factors such as the atmospheric oxygen, CO2, temperature, humidity, and even pH changes which eventually encourages the growth of several microorganisms on the food. The metabolic activities of these microbes can contribute to food poisoning, which can lead to death. The use of packaging (active packaging) materials that relay instantaneous information to consumers about the present status of the packaged foods has been reported by Labuza and Breene (1989) to meet the consumer's demand and increase their confidence level in the products. The use of active packages does not only prevent contamination, today but they are also designed to retain both the quality, organoleptic properties and extend the shelf-life of packaged goods. This type of intelligent packaging also gives updated information on the condition of the product during transportation, storage and eventually during consumption (Ahvenainen, 2003). In the food industry, the packaging of animal-based foods has a lot of beneficial advantages outside boosting the confidence of the eventual consumers of the packaged product. The other benefits include physical and barrier protection, prevention of contamination, provision of product and marketing information, ensured security and good convenience in rationing.

6.6 PACKAGE INFORMATION AND MARKETING

Smart packaging of animal-based food products encourages proper information and labels that provide information, such as the condition of transportation, stage conditions, mode of disposal, the mode of transportation, condition of storage and labels usually come with interesting information such as the food properties which ends up encouraging the consumers to buy the products. The packaging design and concepts also come in colors and designs that are enticing and attractive, and increase public acceptance. Sometimes, these packages come with interesting cartoons that attract kids and provide emotional attachments to the respective products.

6.7 PHYSICAL PROTECTION AND BARRIER PROTECTION

Intelligent, or smart, packaging is done to increase the acceptability of the respective packaged products through several means such as protecting the packaged product from shock or microbial contamination. Another packaging benefit is the protection of the product from water contact, oxygen and finally environmental components that can lead to early deterioration of the packaged food. The self-life of the packaged foods is ensured by using packages that prevent ethylene, oxygen or dust exposure to the product. Also in other to protect the packaged foods, foods that are affected by shock or vibration are designed with the packages to so ensure.

6.8 PREVENTION OF CONTAINMENT

This is done in order to prevent packaged foods from being contaminated by liquid, environmental hazards, and other potential sources.

6.9 SECURITY

Engineering food packages play several roles in ensuring that the organoleptic properties, physicochemical properties, nutritive values, maintain product quality and minimize damages due to transportation across the value or supply chain. These can also be used to provide adequate security for the packaged product. This result will greatly increase the acceptability of the product. During production, the use of RFID tags, or barcodes, ensure proper traceability of the product along the chain of supply, which will give real-time updates of the product information. This is also engaged to ensure that the product is secured from counterfeit or substandard products. These tags are also deployed to monitor the items while in transit.

6.10 CONVENIENCE AND DISPENSING CONTROL

In improving the acceptability of consumers for packaged goods, they must be packaged with materials that ensure easy transportation and distribution without compromising quality, while also enabling the use and reusability of the product. These packaging materials should also be easy to dispense – not too bulky so that the dispensing can easily be controlled. This will protect the repackaged food from accelerated spoilage. Despite observing the active packaging concept, the packages should be easy to administer and reclose/repackage without the packaging material being developed to a high degree of complexity

6.11 FACTORS TO CONSIDER IN DEVELOPING ECONOMICALLY VIABLE PACKAGING SYSTEM MOISTURE AND ENVIRONMENTAL CONTROL

Some of the environmental parameters that affect food spoilage include pH, temperature, moisture content, and oxygen level among others. Control of high moisture content is very important in ensuring the quality and shelf life of any package of preserved meat. Water aids the growth of food spoilage microorganisms and its control in packaged products help extend the shelf life of the product. In recent times, the use of hygroscopic materials/substances absorbs water from the environment. This is designed in form of films and trays for controlling water present in packaged foods. The use of fructose in the membrane of the packaging material is fast emerging in the food industry. Other moisture absorbers currently being used include Sorbitol, potassium chloride, sodium chloride, xylitol, and calcium chloride absorbent resin (Gaikwad et al., 2019).

Aside from moisture absorbers, another important factor being considered in food packaging is the control of excess oxygen. This is usually done with the aid of oxygen scavengers or absorbers in animal-based control. The use of oxygen scavengers inhibits the growth of molds and other microorganisms, resulting in the prevention of food spoilage. Use of sachets containing powdered iron rusts in packaging materials for oxygen removal has been effective but has its own downside due to toxicity. The use of nitrogen in cheese packages has been helpful in increasing shelf-life and preventing spoilage. In ensuring the quality of packaged plant-based foods, temperature sensors and pH sensors are today being installed in packaging materials. Another use of temperature control for animal-based foods involves the use of data inscriptions which are always written on the labels of the materials, and this provides information to transporters, consumers, and even retailers on the best temperature for storage of the product. Preventing food spoilage and contamination through the use of engineered packaged films through designing new packages containing scavengers (CO_2, O_2), antimicrobial agents and sensors that detect color changes, color development, temperature change, moisture level, excess oxygen, corrosion inhibitors and enzyme production due to metabolic processes (Vilas et al., 2020).

6.12 METAL CHELATION

General concern for global health and major industries involves reducing the fast-food deterioration rate which is a result of metal-ion induced oxidation reactions that occur in these foods (Tian et al., 2012). In an attempt to meet the consumers' and global demand for clean foods while preserving the organoleptic and functional food properties of the respective packaged food products, the use of active packaging techniques will result in an increase in the stability of the packaged product by scavenging EDTA or other transition metals present in the packaged product which, when introduced into the package by either covalent immobilization (Muriel-Galet et al., 2013) or roll to roll fabrication, laminated photo-grafting coatings (Lin and Goddard, 2018) improving the surface of the packaging material mostly Polypropylene films by crosslinking agents (Tian et al. 2012) or initiating graft polymerization of acrylic acid by photo effect (light) (Roman et al. 2014). Synthesized food packaging materials made up of polypropylene films with functionalized iminodiacetate films (Lin et al. 2016; Lin et al., 2018), use active packaging materials that chelates iron by biomimetic (Tian et al. 2013) and copolymerization of PE films by grafting that is induced by radiation (Nasef et al. 2021).

6.13 SECURITY

The security of the packaged foods is as important as their quality and safety. This need for security is a result of the presence of sub-standard products in the marketplace today. Packaging materials for animal-based products should ensure they provides security for the product during transportation to ensure proper traceability. The use of radio-frequency identification chips (RFID), which are integrated into packaging labels, makes the product traceable across the chain of supply. It has been established that the application of barcodes that contain detailed information such as production manufacture date, expiry date, location of production, sometimes a price tag, nutritive content, preservation and preparation procedure, and many others. Another new method of ensuring safety of packaged products is the use of QR codes, which reveal details of the packaged food product.

6.14 QR CODES

One of the ways of communicating details of a particular product and product traceability to the consumer about the product is using QR codes. These codes are actually machine labels that are used in either understanding the content or benefits of a product, or they are used in tracking basic information about the product. Specific information such as product transportation, expiration periods, preservation details, and prices are among the things a QR code can provide. In accessing these codes, digital printers or smartphones with scanners can be utilized. These codes are interesting improvements in the food package. Despite the advantages of these codes, they are easy to duplicate and also to counterfeit. Though newer measures, such as the use of secured graphics, sensitive restrictors and digital watermarks are now being incorporated to make them copy sensitive and further strengthen their security and also prevent them from being copied, thus making them reliable. Another way is using the GSI links, which are digital links that provide unique identification for a particular product and ensure labelling of packaged products, which makes them traceable from the production line to the consumer. When providing information to the consumers across retailers and supply chains, unique identifiers are used. This allows the same GSI digital link to be embedded in QR Code.

6.15 ADVANCES IN FOOD PACKAGING

In recent years, an effort to improve food packaging techniques in order to ensure consumers have value for the products they purchase has been yielding positive results recently. The use of special packages for long-distance transportation of organs such as the heart, middle ear, kidney, pancreas, lungs and liver for transplantation has ensured organs are transported alive and fresh to their respective designations. This has increased success rates and reduced complications in the recipient.

Also, the invention of safe sanitizers such as chlorisanitisers in small quantities, especially as small pouches applied in packaging fruits, has largely helped in preventing pathogenic microbes from infecting the products. Other improved technologies in food packaging involve packaging materials that keep track and monitor the condition of the food being packaged right from the production line to the quality assurance stage, boxing, transportation, distribution and eventually to the end-user. Packing of animal-based products with coatings such as chitosan nanoparticle, beeswax, alginate or edible films such as rice paper and sausage casings, which protect packaged foods from external influences, are now on the rise for biotechnological applications in food-based industries.

6.16 REGULATIONS

Since the success of many food industries depends not only on the advertisement strategies but also on packaging techniques which in the long or short-run affect the organoleptic properties of the packaged foods and, finally, their general acceptability, Food industries ensure compliance with at least the minimum standard for smart packaging; there needs to be active regulation(s) for the various packing types used in the animal-based food industries to further enshrine confidence in the consumers who patronize these products. Food packages that are toxic or result in a reduction in the shelf-life of their packaged products are usually avoided. Also, government agencies in several countries all around the world ensure strict compliance to their nations' respective enacted laws to ensure adequate food preservation when packaged.

6.17 CONCLUSION

Antimicrobial control is essential in maintaining the high quality and safety of packaged animal-based food products. The application of diverse packaging techniques in various animal product-based industries is aimed at increasing consumers' confidence in the various packaged foods because animal-based products have a high deterioration rate, and hence, are prone to foodborne pathogens which eventually have detrimental effects on the health of consumers at large. The use of active packaging that provides real-time information to the producers and consumers on the status of the packaged food across the supply chain down to the consumers further increases the general acceptability of these products. The use of barcodes, QR codes and RFIDs further ensures adequate security for the products and reduces the chances of counterfeit products. The application of humidity monitors and sensors in active packaging for temperature, pH and moisture control has resulted in several innovations, such as the addition of moisture, CO2 and O2 scavengers in the lining of the respective packaging membrane. This has further increased the safety level and retained the quality of these packaged products. Even though the times are changing and newer innovations and approaches are being introduced to make animal-based products better packaged, their primary concern should always be ensuring the safety, retaining quality, and the nutritive and functional properties of the packaged foods.

REFERENCES

Adetunji, C.O., Fawole, O.B., Arowora, K.A., Nwaubani, S.I., Ajayi, E.S., Oloke, J.K., Aina, J.A., Adetunji, J.B., and Ajani, A.O. (2013). Postharvest quality and safety maintenance of the physical properties of Daucus *carota L.* fruits by Neem oil and Moringa oil treatment: A new edible coatings. *Agrosearch.* 13(1):131–141.

Adetunji, C.O., Michael, O.S., Nwankwo, W., Ukhurebor, K.E., Anani, O.A., Oloke, J.K., Varma, A., Kadiri, O., Jain, A., and Adetunji, J.B. (2021a). Quinoa, the next biotech plant: Food security and environmental and health hot spots. In: Varma, A. (eds.), *Biology and Biotechnology of Quinoa*. Springer, Singapore. https://doi.org/10.1007/978-981-16-3832-9_19

Adetunji C.O., Michael, O.S., Varma, A., Oloke, J.K., Kadiri, O., Akram, M., Bodunrinde, R.E., Imtiaz, A., Adetunji, J.B., Shahzad, K., Jain, A., Ubi, B.E., Majeed, N., Ozolua, P., and Olisaka, F.N. (2021b). Recent advances in the application of biotechnology for improving the production of secondary metabolites from Quinoa. In: Varma, A. (ed.) *Biology and Biotechnology of Quinoa*. Springer, Singapore. https://doi.org/10.1007/978-981-16-3832-9_17

Adetunji, C.O., Olugbenga, S.M., Kadiri, O., Varma, A., Akram, M., Oloke, J. K., Shafique, H., Adetunji, J. B., Jain, A., Bodunrinde, R. E., Ozolua, P., and Ubi, B. E. (2021c). Quinoa: From farm to traditional healing, food application, and phytopharmacology. In: Varma A. (ed.), *Biology and Biotechnology of Quinoa*. Springer, Singapore. https://doi.org/10.1007/978-981-16-3832-9_20

Adetunji, C.O., Olaniyan, O.T., Adeyomoye, O., Dare, A., Adeniyi, M.J., Alex, E., Rebezov, M., Garipova, L., and Ali Shariati, M. (2022a). eHealth, mHealth, and telemedicine for COVID-19 pandemic. In: Pani, S.K., Dash, S., dos Santos, W.P., Chan Bukhari, S.A., & Flammini, F. (eds), *Assessing COVID-19 and Other Pandemics and Epidemics using Computational Modelling and Data Analysis*. Springer, Cham. https://doi.org/10.1007/978-3-030-79753-9_10

Adetunji, C.O., Olaniyan, O.T., Adeyomoye, O., Dare, A., Adeniyi, M.J., Alex, E., Rebezov, M., Petukhova, E., and Ali Shariati, M. (2022b). Machine Learning approaches for COVID-19 pandemic. In: Pani, S.K., Dash, S., dos Santos, W.P., Chan Bukhari, S.A., Flammini, F. (eds). *Assessing COVID-19 and Other Pandemics and Epidemics using Computational Modelling and Data Analysis*. Springer, Cham. https://doi.org/10.1007/978-3-030-79753-9_8

Adetunji, C. O., Olaniyan, O.T., Adeyomoye, O., Dare, A., Adeniyi, M.J., Alex, E., Rebezov, M., Isabekova, O., and Ali Shariati, M. (2022c). Smart sensing for COVID-19 pandemic. In: Pani, S.K., Dash, S., dos Santos, W.P., Chan Bukhari, S.A., Flammini, F. (eds), *Assessing COVID-19 and Other Pandemics and Epidemics using Computational Modelling and Data Analysis*. Springer, Cham. https://doi.org/10.1007/978-3-030-79753-9_9

Adetunji, C.O., Olaniyan, O.T., Adeyomoye, O., Dare, A., Adeniyi, M.J., Alex, E., Rebezov, M., Petukhova, E., and Ali Shariati, M. (2022d). Internet of Health Things (IoHT) for COVID-19. In: Pani, S.K., Dash, S., dos Santos, W.P., Chan Bukhari, S.A., Flammini, F. (eds). *Assessing COVID-19 and Other Pandemics and Epidemics using Computational Modelling and Data Analysis*. Springer, Cham. https://doi.org/10.1007/978-3-030-79753-9_5

Adetunji, C.O., Olaniyan, O.T., Adeyomoye, O., Dare, A., Adeniyi, M.J., Alex, E., Rebezov, M., Koriagina, N., and Ali Shariati, M. (2022e). Diverse techniques applied for effective diagnosis of COVID-19. In: Pani S.K., Dash S., dos Santos W.P., Chan Bukhari S.A., Flammini F. (eds), *Assessing COVID-19 and Other Pandemics and Epidemics using Computational Modelling and Data Analysis*. Springer, Cham. https://doi.org/10.1007/978-3-030-79753-9_3

Adetunji, C. O., Ogundolie, F. A., Ajiboye, M. D., Mathew, J. T., Inobeme, A., Dauda, W. P., and Adetunji, J. B. (2022f). Nano-engineered sensors for food processing. In Bio- and Nano-sensing Technologies for Food Processing and Packaging, pp. 151–166. *Royal Society of Chemistry*. https://doi.ord/10.1039/9781839167966-00151

Adetunji, C. O., Mathew, J. T., Inobeme, A., Olaniyan, O. T., Singh, K. R. B., Abimbola, O. F., Nayak, V., Singh, J. and Singh, R. P. (2022g). Microbial and plant cell biosensors for environmental monitoring. In: Singh, R. P., Ukhurebor, K. E., Singh, J., Adetunji, C. O., and Singh, K. R. (eds.), *Nanobiosensors for Environmental Monitoring*. Springer, Cham. https://doi.org/10.1007/978-3-031-16106-3_9

Ahvenainen, R. (ed.). (2003). *Novel Food Packaging Techniques*. Elsevier. Woodhead Publishing: Cambridge.

Arvanitoyannis, I. S. (2005). Food packaging technology. In: R. Coles, D. McDowell, and M. J. Kirwan. Blackwell Publishing, CRC Press, Oxford, 2003. In *Journal of the Science of Food and Agriculture* 85(6): 1072. doi:10.1002/jsfa.2089

Bi, L. J. (2012). Research on corrugated cardboard and its application. *Advanced Materials Research*, 535–537: 2171–2176. doi:10.4028/www.scientific.net/AMR.535-537.2171.

Gaikwad, K. K., Singh, S., and Ajji, A. (2019). Moisture absorbers for food packaging applications. *Environmental Chemistry Letters*, 17(2): 609–628.

Han, J. W., Ruiz- Garcia, L., Qian, J. P., and Yang, X. T. (2018). Food packaging: A comprehensive review and future trends. *Comprehensive Reviews in Food Science and Food Safety*. Doi:10.1111/1541-4337.12343

Hauchhum, L., and Mahanta, P. (2014). Carbon dioxide adsorption on zeolites and activated carbon by pressure swing adsorption in a fixed bed. *International Journal of Energy and Environmental Engineering*, 5: 349–356.

Kerry, J. and Butler, P. (2008). Smart Packaging Technologies for Fast Moving Consumer Goods. *Smart Packaging of Meat and Poultry Products*. 10.1002/9780470753699(), 33–59. doi:10.1002/9780470753699.ch3

Kerry, J. P., O'Grady, M. N., and Hogan, S. A. (2006). Past, current and potential utilisation of active and intelligent packaging systems for meat and muscle-based products: A review. *Meat Science*, 74(1): 0–130. doi:10.1016/j.meatsci.2006.04.024

Kuswandi, B. (2017). Reference module in food science. *Freshness Sensors for Food Packaging*, 7(2): 1–11. doi:10.1016/B978-0-08-100596-5.21876-3.

Lee, D. S. (2016) Carbon dioxide absorbers for food packaging applications. *Trends in Food Science & Technology*, 57: 146–155. https://doi.org/10.1016/j.tifs.2016.09.014

Lin, Z., Roman, M. J., Decker, E. A., and Goddard, J. M. (2016). Synthesis of iminodiacetate functionalized polypropylene films and their efficacy as antioxidant active-packaging materials. *Journal of Agricultural and Food Chemistry*, 64(22): 4606–4617. https://doi.org/10.1021/acs.jafc.6b01128

Lin, Z. and Goddard, J. (2018). Photo-curable metal-chelating coatings offer a scalable approach to production of antioxidant active packaging. *Journal of Food Science*, 83(2): 367–376. https://doi.org/10.1111/1750-3841.14051

Mathew, R. and Jaganathan, D. (2017). Packaging and storage practices of meat. *Global Journal of Biology Agricultural and Health Science*, 6(1): 32–40, doi:10.24105/Gjbahs.6.1.1707.

Mehrotra, P. (2016). Biosensors and their applications – A review. *Journal of Oral Biology and Craniofacial Research*, 6(2): 153–159. https://doi.org/10.1016/j.jobcr.2015.12.002

Mohebi, E., and Marquez, L. (2015). Intelligent packaging in meat industry: An overview of existing solutions. *Journal of Food Science and Technology*, 52(7): 3947–3964. https://doi.org/10.1007/s13197-014-1588-z

Muriel-Galet, V., Talbert, J. N., Hernandez-Munoz, P., Gavara, R., and Goddard, J. M. (2013). Covalent immobilization of lysozyme on ethylene vinyl alcohol films for nonmigrating antimicrobial packaging applications. *Journal of Agricultural and Food Chemistry*, 61(27): 6720–6727. https://doi.org/10.1021/jf401818u

Nasef, M. M., Gupta, B., Shameli, K., Verma, C., Ali, R. R., and Ting, T. M. (2021). Engineered bioactive polymeric surfaces by radiation induced graft copolymerization: Strategies and applications. *Polymers*, 13(18): 3102. https://doi.org/10.3390/polym13183102

Olayemi, F. F. and Adetunji, C. O. (2013). Effect of rinses on microbial quality of commercially available eggs and its components before processing from Ilorin in western Nigeria. *Bitlis Eren University Journal of Science and Technology*, 3(2): 44–47. Bitlis Eren University, Turkey.

Ogundolie, F. A., Ayodeji, A. O., Olajuyigbe, F. M., Kolawole, A. O., and Ajele, J. O. (2022). Biochemical Insights into the functionality of a novel thermostable β-amylase from *Dioclea reflexa. Biocatalysis and Agricultural Biotechnology*, 42: 102361. https://doi.org/10.1016/j.bcab.2022.102361

Opie, R. *Packaging Source Book. Macdonald Orbis*. OCLC 19776457.

Puligundla, P., Jung, J., and Ko, S. (2012). Carbon dioxide sensors for intelligent food packaging applications. *Food Control*, 25: 328–333.

Risch, Sara J. (2009). Food packaging history and innovations. *Journal of Agricultural and Food Chemistry*. 57 (18): 8089–8092. doi:10.1021/jf900040r. ISSN 0021-8561. PMID 19719135.

Robertson, G. (2013). *Food Packaging Principles and Practices*, third ed. Taylor and Francis Group, Boca Raton, FL.

Roman, M. J., Decker, E. A., and Goddard, J. M. (2014). Metal-chelating active packaging film enhances lysozyme inhibition of Listeria monocytogenes. *Journal of Food Protection*, 77(7): 1153–1160. https://doi.org/10.4315/0362-028X.JFP-13-545

Saha, B. B., Jribi, S., Koyama, S., and El-Sharkawy, I. (2011). Carbon dioxide adsorption isotherms on activated carbons. *Journal of Chemical and Engineering Data*, 56: 1974–1981.

Smith, J. D., Dhiman, R., Anand, S., Reza-Garduno, E., Cohen, R.E.; McKinley, G.H., and Varanasi, K.K. (2013). Droplet mobility on lubricant-impregnated surfaces. *Soft Matter* 9(6): 1772–1780.

Soroka, W. (2008). Illustrated glossary of packaging terms. *Institute of Packaging Professionals*. p. 3.

Thakur, M. S., and Ragavan, K. V. (2013). Biosensors in food processing. *Journal of Food Science and Technology*, 50(4): 625–641. https://doi.org/10.1007/s13197-012-0783-z

Tian, F., Decker, E. A., and Goddard, J. M. (2012). Development of an iron chelating polyethylene film for active packaging applications. *Journal of Agricultural and Food Chemistry*, 60(8): 2046–2052.

Tian, F., Decker, E. A., and Goddard, J. M. (2013). Controlling lipid oxidation via a biomimetic iron chelating active packaging material. *Journal of Agricultural and Food Chemistry*, 61(50): 12397–12404. https://doi.org/10.1021/jf4041832

Vilas, C. and Mauricio-Iglesias, G. (2020). Model-based design of smart active packaging systems with antimicrobial activity. *Food Packaging and Shelf Life*. 24: 100446. doi:10.1016/j.fpsl.2019.100446. hdl:10347/20889

Yildirim, S., Röcker, B., Pettersen, M.K., Nilsen-Nygaard, J., Ayhan, Z., Rutkaite, R., Radusin, T., Suminska, P., Marcos, B., and Coma, V. (2017). Active packaging applications for food. *Comprehensive Reviews in Food Science and Food Safety*. doi:10.1111/1541-4337.12322

7 Application of Gas/ Biosensors for Effective Determination of Contaminants in the Environment and Food Products as Well as Their Role in Detection of Toxic Pesticides

Abel Inobeme, Charles Oluwaseun Adetunji, Akinola Samson Olayinka, and Olalekan Akinbo

CONTENTS

7.1 INTRODUCTION

The worldwide population has increased swiftly and is expected to increase more in the near future. This has prompted the massive use of various classes of pesticides with a view to meeting the global demand for food through boosting of agricultural yield. Pesticide involves a naturally derived or synthetic substance or combination of compounds that regulate, or act, on the mechanism of functioning of the body of an organism that constitutes a threat to crops, animals and overall agricultural yield as well as the domestic ecosystem (EPA, 2016). International Union of Pure and Applied Chemistry (IUPAC)defined pesticides as compounds effective in preventing,

DOI: 10.1201/9781003207955-7

killing or combating undesirable species that can in any way influence food security, manufacturing, storage and transport, farm produce generally, animal products and all other agricultural outputs (Stephenson et al., 2006). Farmers use various pesticides in farming to protect crops and plants, prior to and after harvesting. The United States governmental department that is concerned with the protection of the environment, the Environmental Protection Agency (EPA), gave the description of pesticide as any chemical or combination of chemical substances that are intended to kill, destroy, deter or modify any disease. It is apparent from these descriptions that the term *pesticide* goes beyond just insecticides but is also applicable to fungicides and herbicides, as well as other chemical compounds that are employed for the combating of agricultural and domestic pests. Although their use is advantageous, such use poses significant risks on both public health and safety (Hameed et al., 2016).

Pesticides are commonly used to involve organic toxic compounds that are for insect control, fungi, plants, nematodes, animals, and other pests. There are different means through which pesticide residues enter the food cycle, eventually coming in contact with humans: through air, soil, water amongst others (Adetunji et al., 2021a,b,c; Ukhurebor et al., 2021; Sangeetha et al., 2021; Adetunji and Ukhurebor, 2021; Oluwaseun et al., 2017; Adetunji et al., 2012; Arowora et al., 2012; Dauda et al., 2022a,b; Ukhurebor et al. 2021; Okeke et al., 2021; Adetunji and Anani, 2021; Nwankwo et al., 2021; Adejumo and Adetunji. 2018; Ukhurebor et al. 2021).

They affect the entire ecosystem significantly, thereby resulting in health-related problems in humans and other animals. Pesticides can be cytotoxic or carcinogenic. They can cause diseases of the blood cells and nerves, infertility, and endocrine and lung diseases (Sassolas et al., 2012).

Owing to their higher insecticidal efficacy, pesticides are extensively employed in agriculture and industry all over the world. Pesticides are poisonous chemicals used for agricultural purposes and accountable for extreme forms of global pollution. They can also bring about deleterious chronic effects and disruption to development. Finding remediation approaches is a big concern regarding environmental and human health effects (Poirier et al., 2017). Constant and unsustainable use of chemical substances has resulted in the destruction of different habitats in different regions of the world. A recent study has revealed that randomly studied areas, including catchments in Scotland and some aquatic environments in Wales, had high levels of contamination from pesticides, most especially from the organophophorus group (Grung et al., 2015).

Soil is also a dynamic and complex ecosystem that plays a major role in production of food and in the sustenance of all other anthropogenic activities on earth. Environmental conservation involves controlling the pollutants to avoid serious impacts. Pesticidal soil pollution poses a hazard to nature, affecting the ecosystem as well as the associated food chains, quality of water and the environment (Muenchen et al., 2016;Adetunji et al., 2020; Adetunji et al., 2020a,b,c; Oyedara et al., 2022; Adetunji et al., 2022a,b,c,d,e,f,g,h,I;Olaniyan et al., 2022).). The presence of chemical contaminants and other toxic substances in water, soil, air and food is presently one of the main environmental chemistry concerns. Pesticides are amongst the most significant contaminants to the environment due to their growing use in agricultural production (Chauhan, et al., 2015).

The widespread use of pesticides, such as the phosphorus-based pesticides like organophosphorus are potent toxins to the nervous system and the subsequent health-related concerns (CDC, 2005). They have generated a need for ways to remove such chemicals from the ecosystem (Stoytcheva, 2010). Due to the high toxicity of many of these pesticides such as organophosphate (OP) and their heavy usage, seeking effective decontamination methods is a big concern. Numerous techniques have been considered for the development of remediation strategies against Ops, including chemical, physical, and biological methods. Nevertheless, such approaches typically require harsh conditions, are not consistent, and cannot be used for environmental remediation on a large scale (Poirier et al., 2017).

Many compounds used as pesticides have been confirmed to be highly poisonous and powerful cholinesterase inhibitors that pose health problems to humans and constitute challenges to the environment at large. A specific group of these commonly used pesticides is composed of organothiophosphates with a functional group of thiophosphoryl (P=S). They are associated with the very volatile organophosphates of phosphoryl (P=O), such as agents used in chemical wars like Sarin and Soman. The wide usage of phosphorus-based organic compounds as pesticide is responsible for the increasing cases of pesticide residues in agricultural produce such as crops, fruits, vegetables and their consequential infiltration into water bodies and aquifers (Obare et al., 2010).

Several kinds of pesticides are utilized by farmers for the purpose of food security, protecting crops and other agricultural plants and grains during pre- and post-harvesting periods. The compounds contaminate our environment and can reach water, soil, and air as well as contaminate the entire food chain. They affect the environment significantly and can cause harm to humans and livestock. Many are also known to cause cancerous and genotoxic consequences in man. These can cause various forms of abnormalities such tumor growth, nerve diseases, depression, immune and respiratory diseases (Hameed et al., 2016).

Amongst the group of pesticides, the chemical pesticide that is most known and commonly used in agriculture are organophosphate-based, such as the malathion, parathion and chlorpyrifos. The tubes on the systems create droplets when pesticide solutions are sprinkled. Thus they continue to suspend in air and are transported to other places by wind currents, eventually infecting areas outside the intended usage sites.

In addition, pesticides can endanger local biodiversity during application to farmlands by escaping into the environment and the action of wind disperses them into distant areas not originally targeted (Kaur et al., 2007). Moreover, given the wide-ranging advantages of using pesticides in agriculture, some wrong applications may lead to high and unacceptable levels of compounds in the product reaching consumers. They involve the incorrect use of synthetic compounds on foodstuffs, including the harvesting of crops before the residues have been washed off after usage (Gill and Garg, 2013).

More than 130 distinct pesticides are being used globally for the processing of food types, with nearly 25 frequently used compounds. They are of different categories such as carbamates, chlorine based, organophosphate, compounds of nitrogen, amides, and heterocyclic compounds pesticides. Some of these compounds are highly toxic, affecting vital organs in humans and animals. They are also capable of affecting the nervous and respiratory systems or acts as carcinogenic and genotoxic agents, thereby raising the health concern connected with polluted foods. Specifically, pesticide such as chlorpyrifos, which is phosphorus based, affect the nervous system (Mavrikou et al., 2008).

Pesticides are classified based on different factors; in some basis of grouping, the targeted pest for which the pesticide is applied is considered, other factors usually employed in the classification include, the extent of toxicity, plant resistance to the pesticide, structural form and source of the pesticide amongst others. The basis of toxicity tables commonly show their extent of dermal and oral toxicity to end users. Such tables are periodically reversed to accommodate changes based on recent studies. According to the purpose they are used for, pesticides are classified into bactericides, for killing bacterials, fungicides for attacking fungi, rodenticides for rat, rabbit, squirrel and related organisms, molluscidcides for snails and other mollusks, and many others.

Pesticides can be administered manually or by contact due to their versatility. The application is to one area of the plant and usually by foliage, so that they can enter other nontreated areas. The World Health Organization (WHO) categorize pesticides into four groups, taking into consideration their toxicity: those that are of extreme toxicity (IA), high toxicity (IB), moderate toxicity (II) and slight toxicity (III) (Zamora-Sequeira et al., 2019).

Biosensors provide a convenient method for assessing and tracking pesticides in atmospheric and food matrices. The use of biosensors as screening instruments is economical and minimizes the number of samples to be tested by the aforementioned conventional laboratory techniques. Wireless technology and sensor technology have become a crucial resource for daily life around the world with the exponential growth of smart phones. The next generation of a hyper-connected planet, the Internet of Things (IoT), reflects the coming wave of interconnected computers, appliances, sensors, meters and countless other "things." Interconnected entities should open a channel of communication between themselves (Bibri et al., 2020).

Biosensors are usually characterized as a selective component that is chemically active, containing a biological component and incorporated with a transducer unit. The biological source may be an antigen, biological protein such as enzyme, an antibody, strands of deoxyribonucleic acids, aptamer, a tissue or even a living organism component. The mechanism of functioning of the transducer can be through its sensitivity to photons, temperature, electron flow or any other physical or chemical parameter.

The Electrode is a transducer that is well known as commonly used. The development of various novel biosensing devices is due to the advancement in area of biotechnology and material science and chemical and biological information interconversion. There are varieties of sensors that are capable of converting the concentration of the measured substance into signals that can be quantified analytically (Stoytcheva, 2010). Biological substances play a very important role in the field of biosensors as one of the key components for measuring the system's selectivity and specificity. The identification mechanism is the basic binding of the specimen of interest to the corresponding biological material immobilized on an acceptable support matrix (Kumar et al., 2018).

Therefore, this chapter provides a general overview of the application of toxic as sensors and biosensors for the degrading toxic pesticides by gas/biosensors. Specific examples of these biosensors that are normally applied in the environmental and food sector are also highlighted.

7.2 RELEVANCE OF TOXIC GAS SENSORS AND BIOSENSORS IN THE DETECTION OF TOXIC PESTICIDES

Boosting production yield for effective enhancement in agriculture is done through the application of synthetic chemicals. The chemicals used in pesticide are however highly toxic and have deleterious consequences when they accumulate in the environment or their residues come into contact with the bodily system. There is therefore a need for their degradation and reliable strategies for monitoring their presence in the environment. One of the most reliable means for checking for the manifestation of toxic pesticide remains in the environment is the use of biosensing devices that function based on an enzyme inhibition mechanism. This approach is far more important due to its promising biological importance and its reliability.

Recent progress has been made with the production of biosensors that target cholinesterase inhibition, phosphatase that have pH greater than 7, photosynthesis based systems, tyrosinase, and dehydrogenase among others. Different classes of chemicals such as carbamates, diazines, phenols and dithiocarbamates have been detected using different techniques. However, some of these approaches have not been widely applied to environment and food issues since tests involving their usage have been highly limited (Bucur et al., 2018).

The basic analytical methods used for pesticide detection are chromatographic techniques, combined with specific detectors. Both methods have the advantages of being automatic and precise, highly sensitive, and can be used to detect simultaneously. However, in spites of the advantages associated with these systems, they also have their specific disadvantages such their expensive nature, time consumption during use, requisition for sample pretreatment test which could be tedious, and the fact that it also requires high technical knowhow, hence properly trained personnel are highly necessary (Bucur et al., 2018). For an analytical method to be considered useful and relevant in this aspect of pesticide detection and monitoring, it must show a high level of sensitivity and precision, also giving rise to responses within the shorter time possible so as to enhance ease of detecting various pesticide traces from a wide variety of environmental samples. The mechanism for pesticides residues that are toxic to the nervous system is usually based on acetylcholinesterase inhibition. This has been utilized in the monitoring of multianalyte systematically through replication in in vitro studies (King and Aaron, 2015).

The effective and reliable pesticide detection tool is the use of the second enzyme, for example AchE and choline oxidase (Cox). Biosensors based on multienzymes use cholinesterase in connection with choline oxidase to monitor the production of hydrogen peroxide or the consumption of oxygen. Cox oxidizes betaine and hydrogen peroxide into choline. This method makes use of two steps that are interrelated and interdependent. First an enzyme that is repressed by the pesticide is coupled with another enzyme that depends on the outcome of the activity of the first enzyme. Studies have also reported another enzyme that works based on amperometricbiosensing in the detection of the presence in the environment of dithiocarbamate, a fungicide (Ju and Kandimalla, 2008).

Electrochemical sensors are useful in the detection of pesticide residues, their detection based on use of acylcholinesterases or phosphates, whether in alkaline or acidic environments, is based on the tendency of the organophosphate compound to block the activity of the enzyme. The amount present is determined using electrochemical techniques to calculate the variability of the enzymatic behavior as the amount of the organophosphate change in the pesticide. The transduction mechanism therefore affects the detection of the pesticide by either potentiometric or amperometric means (Stoytcheva, 2010).

It has been reported based on findings by Poirier et al. (2017) that the enzyme known as organophosphorus hydrolase functions by the hydrolyzing organophophotriester; the enzyme is capable of acting on different phosphorus containing pesticides such as coumaphos, diazinon, parathion amongst others. The anhydrolase of organophosphorus acid functions by attacking bonds such as phosphorus/cyano group, phosphorus/oxygen, phosphorus/sulphur, thereby bringing about the catalysis of the hydrolysis of the organophosphorus compounds through the generation of two protons. The hydrogen ion generated could then be followed by potentiometric means. It is possible to monitor the reduction and oxidation current of the products of hydrolysis by the incorporation of organophosphorus hydrolase with a transducer that is suitable for amperometry.

Using this approach the toxic effect of parthaion ethyl, raraoxon ethyl and fenitrothion, which are commonly available and used pesticides, were assessed. The specific products of oxidation generated by enzymatic hydrolysis were also assessed. In *Schmidteamediterranea*, mortality, behavior, and regeneration were assessed. It has been shown that organophosphate is toxic to plenaries and affects their ability to move and regenerate in cases of injuries, however when certain enzymes are employed, the harmful effects of the pesticide compounds are reduced significantly.

Chauhan et al. (2012) were able to innovate a novel biosensing mechanism for the identification of synthetic compounds in the environment established on electrochemical principles. This involved the bioconjugation of of acetylcholinesterase with iron oxide nanoparticles and poly(indole-5-carboxylic acid). The electrochemical sensor developed was high in its precision and accuracy due to the presence of iron oxide nanocomposite as a sensing interphase highly amplified at the lower voltage.

The detection of the enzyme that inhibits the pesticide activity was done under a neutral pH and room temperature with a concentration varying from 0.1 to 60nM for malathion and 1.5 to 70 nM for clorpyrifos. In a similar study, an extremely sensitive biosensing device was established by Chauhan, et al. (2015) for the quantitative determination of pesticide residues.

Hondred et al. (2018) in their study detected the presence of organophosphate compounds using an electrochemical biosensor with a high sensitivity and speed designed by Maskless Inkject Lithrography. In the design, the organophosphate enzyme, which is hydrolase, was joined by conjugation to a layer of electrode by cross-linking with gluraraldehyde. The biosensor developed was used accurately for the quantification of paraoxone an organophosphate-based insecticide with a sensitivity of 370nA/uM and a minimum amount of interferences.

In addition, the biosensor demonstrated high reusability (average reduction in sensitivity per sensing event of 0.3 percent), durability (90 percent anodic current pulse retention over 1,000 seconds), endurance (the sensitivity was retained even after a period of eight weeks), and had the ability of detecting even trace amounts of the pesticide in different environmental matrices.

7.3 APPLICATION CHEMICAL SENSOR FOR THE DETECTION OF CONTAMINANTS PRESENT IN FOOD

The application of electronic nose is an effective and sustainable diagnostic tool for effective identification of high level of contamination and pathogens that are present on food. Chemical sensor systems have been used in the food sector for the identification of contaminants, most especially toxins, and numerous foodborne pathogens. The application of odor-mapping approaches has been documented toward the determination of grains quality which is normally carried out by human olfactory panels through the investigative officers who carried out the smelling of the odor from the grains. Figure 7.1 shows the application of sensor technology in handling different contaminations.

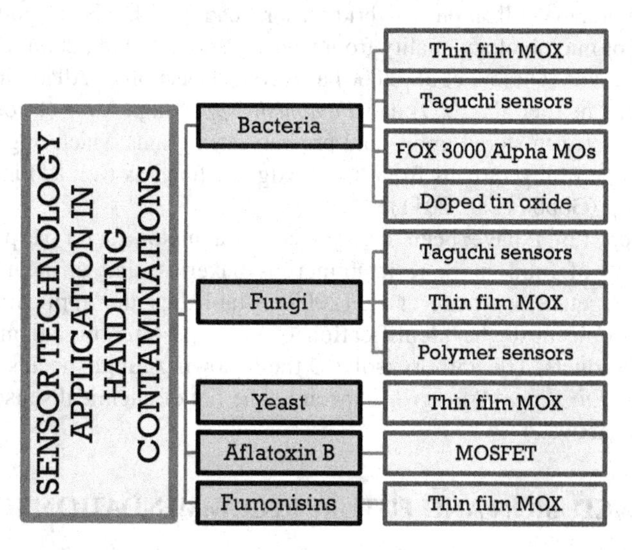

FIGURE 7.1 Sensor Technology Application in Handling Contaminations.

TABLE 7.1
Utilization of Chemical Sensor in Food Sector

Types of food	Food based Product	Types of contamination	Sensor technology	References
Beverages and Drinks	Red wine	Bacteria (*Alicyclobacillus*)	FOX 3000 Alpha MOs	Concina et al., 2010
Grains and cereals	Oats rye and barley	Bacteria and fungi	MOSFET and Taguchi	Jonssonet al., 1997
	Maize	Fumonisins and fungi	Thin film MOX	Falasconi et al., 2005 Gobbi et al., 2011
	Barley	fungi	Taguchi sensors	Olsson et al, 2000
Meat	Sheep meat and beef	Bacteria	Taguchi sensors	Barbri et al., 2008
Fruits and vegetables	Tomatoes	Yeast, bacteria, fungi	Thin-film MOX	Concina et al., 2009
	Onion	Bacteria and fungi	Polymer sensors	Li et al., 2011
Milk product	Milk derived from ewes	Aflatoxin B	MOSFET	Benedetti al., 2005
	Milk	Bacteria	MOX	Labreche et al., 2005
	Sardines	Bacteria	Doped tin oxide	Barbri et al., 2009

The application of electronic nose represents a sustainable approach because of its effectiveness, economic value, simplicity and high level of sensitivity. Moreover, this will go a long way towards mitigating against several hazards that the investigator or panelists could be exposed to, most especially food contaminants such as mycotoxin and molds.

Furthermore, the utilization of hybrid sensor technology has been validated for the assessment of microbiology quality from wheat (Paolesse, et al., 2006; Evans, et al., 2000), and grains (Olsson, et al., 2000) as well as detection of Aflatoxin (Paolesse, et al., 2006; Cheli, et al., 2009) and *Fumonisin spp.* Ahigh level of contamination has been detected in stored grains and pre-harvested grains which might be linked to the presence of mycotoxins from mycotoxigenic fungi as well as fumonisins, by *Fusarium spp* (Gobbi, et al., 2011).

Moreover, fungi have been recognized as a major factor responsible for the high rate of food spoilage in numerous bakeries, and in meat as well as fruits and vegetables. Marín et al. (2007) established the application of MS-based electronic nose the identification of spoilage microorganisms available in bakery products. The authors isolated the following fungi species such as the *Penicillium, Eurotium, Aspergillus* species. The list of chemical sensor is shown in Table 7.1 above.

7.4 CONCLUSION AND FUTURE RECOMMENDATION TO STUDY

This chapter has provided detailed information on the application of toxic gas sensors and biosensors for degrading toxic pesticides by gas/biosensors. Detailed information is given on the application of organophosphorus hydrolasea typical example of how the enzyme is capable of acting on different phosphorus-containing pesticides such as coumaphos, diazinon, parathion, amongst others. The modes of action through which these enzymes attack bonds such as phosphorus/cyano group, phosphorus/oxygen, phosphorus/sulphur, leading to the catalysis of the hydrolysis of the organophosphorus compound were also highlighted.

REFERENCES

Adetunji CO, Anani, OA (2021). Bioaugmentation: A Powerful Biotechnological Techniques for Sustainable Ecorestoration of Soil and Groundwater Contaminants. In: Panpatte, DG, Jhala YK (eds) *Microbial Rejuvenation of Polluted Environment. Microorganisms for Sustainability*, vol 25. Springer, Singapore. https://doi.org/10.1007/978-981-15-7447-4_15

Adetunji CO, Fawole OB, Afolayan SS, Olaleye OO, Adetunji JB (2012). *An oral presentation during 3rd NISFT western chapter half year conference/general meeting*. Ilorin. pp. 14–16.

Adetunji CO, Ukhurebor KE (2021). Recent Trends in Utilization of Biotechnological Tools for Environmental Sustainability. In: Adetunji CO, Panpatte DG, Jhala YK (eds) *Microbial Rejuvenation of Polluted Environment. Microorganisms for Sustainability*, vol 27. Springer, Singapore. https://doi.org/10.1007/978-981-15-7459-7_11

Adetunji, CO, Mitembo WP, Egbuna C, Narasimha Rao GM (2020). Silico Modeling as a Tool to Predict and Characterize Plant Toxicity. In: Andrew G. Mtewa, Chukwuebuka Egbuna, Narasimha Rao GM (eds) *Poisonous Plants and Phytochemicals in Drug Discovery*. https://doi.org/10.1002/9781119650034. Published by Wiley Online Library.

Adetunji CO, Samuel MO, Nwankwo W, Ukhurebor KE, Anthony Anani O, Oloke JK, Varma A, Kadiri O, Jain A, Adetunji JB (2021a). Quinoa, The Next Biotech Plant: Food Security and Environmental and Health Hot Spots. In: Varma, A. (eds) *Biology and Biotechnology of Quinoa*. Springer, Singapore. https://doi.org/10.1007/978-981-16-3832-9_19

Adetunji CO, Osikemekha Anthony Anani, Olaniyan OT, Inobeme A, Olisaka FN, Uwadiae EO, Obayagbona ON (2021b). Recent Trends in Organic Farming. In: Soni, R., Suyal, D.C., Bhargava, P., Goel, R. (eds) *Microbiological Activity for Soil and Plant Health Management*. Springer, Singapore. https://doi.org/10.1007/978-981-16-2922-8_20

Adetunji CO, Egbuna C, Oladosun TO, Akram M, Michael O, Olisaka FN, Ozolua P, Adetunji JB, Enoyoze GE, Olaniyan O. (2021c). Efficacy of Phytochemicals of Medicinal Plants for the Treatment of Human Echinococcosis. In: *Neglected Tropical Diseases and Phytochemicals in Drug Discovery*. Ch 8. DOI: 10.1002/9781119617143

Adetunji CO, Olaniyan OT, Adeyomoye O, Dare A, Adeniyi MJ, Alex E, Rebezov M, Garipova L, Shariati MA. (2022a). eHealth, mHealth, and Telemedicine for COVID-19 Pandemic. In: Pani S.K., Dash S., dos Santos W.P., Chan Bukhari S.A., Flammini F. (eds.). *Assessing COVID-19 and Other Pandemics and Epidemics using Computational Modelling and Data Analysis*. Springer, Cham. https://doi.org/10.1007/978-3-030-79753-9_10

Adetunji CO, Olaniyan OT, Adeyomoye O, Dare A, Adeniyi MJ, Alex E, Rebezov M, Petukhova E, Shariati MA. (2022b). Machine Learning Approaches for COVID-19 Pandemic. In: Pani S. K., Dash S., dos Santos W. P., Chan Bukhari, S.A., Flammini, F. (eds.). *Assessing COVID-19 and Other Pandemics and Epidemics using Computational Modelling and Data Analysis*. Springer, Cham. https://doi.org/10.1007/978-3-030-79753-9_8

Adetunji, CO, Olaniyan OT, Adeyomoye O, Dare A, Adeniyi MJ, Alex E, Rebezov M, Petukhova E, Shariati MA.. (2022c). Smart Sensing for COVID-19 Pandemic. In: Pani S.K., Dash S., dos Santos W. P., Chan Bukhari S. A., Flammini F. (eds.). *Assessing COVID-19 and Other Pandemics and Epidemics using Computational Modelling and Data Analysis*. Springer, Cham. https://doi.org/10.1007/978-3-030-79753-9_9

Adetunji, CO, Olaniyan OT, Adeyomoye O, Dare A, Adeniyi MJ, Alex E, Rebezov M, Petukhova E, Shariati MA. (2022d). Internet of Health Things (IoHT) for COVID-19. In: Pani S. K., Dash S., dos Santos W. P., Chan Bukhari S. A., Flammini F. (eds.). *Assessing COVID-19 and Other Pandemics and Epidemics using Computational Modelling and Data Analysis*. Springer, Cham. https://doi.org/10.1007/978-3-030-79753-9_5

Adetunji, CO, Olaniyan OT, Adeyomoye O, Dare A, Adeniyi MJ, Alex E, Rebezov M, Petukhova E, Shariati MA. (2022e). Diverse Techniques Applied for Effective Diagnosis of COVID-19. In: Pani, S.K., Dash, S., dos Santos, W.P., Chan Bukhari, S.A., Flammini, F. (eds) *Assessing COVID-19 and Other Pandemics and Epidemics using Computational Modelling and Data Analysis*. Springer, Cham. https://doi.org/10.1007/978-3-030-79753-9_3

Adetunji, CO, Olaniyan OT, Adeyomoye O, Dare A, Adeniyi MJ, Alex E, Rebezov M, Petukhova E, Shariati MA (2022f). Antiprotozoal Activity of Some Medicinal Plants against Entamoeba Histolytica, the Causative Agent of Amoebiasis. In: *Medical Biotechnology, Biopharmaceutics, Forensic Science and Bioinformatics*. CRC Press. Pages 12. eBook. www.taylorfrancis.com/chapters/edit/10.1201/9781003178903-20/antiprotozoal-activity-medicinal-plants-entamoeba-histolytica-causative-agent-amoebiasis-charles-oluwaseun-adetunji-oyetunde-oyeyemi

Adetunji, CO, Nwankwo W, Olayinka AS, Olugbemi OT, Akram M, Laila U, Samuel MO, Oshinjo AM, Adetunji JB, Okotie GE, Esiobu N. (2022g). Computational Intelligence Techniques for Combating COVID-19. In: *Medical Biotechnology, Biopharmaceutics, Forensic Science and Bioinformatics*. CRC Press. Pages 12. eBook. DOI: 10.1201/9781003178903-16

Adetunji CO, Olugbemi OT, Akram M, Laila U, Samuel MO, Oshinjo AM, Adetunji JB, Okotie GE, Esiobu N, Oyedara OO, Adeyemi FM. (2022h). Application of Computational and Bioinformatics Techniques in Drug Repurposing for Effective Development of Potential Drug Candidate for the Management of COVID-19. In: *Medical Biotechnology, Biopharmaceutics, Forensic Science and Bioinformatics*. CRC Press. P. 14. eBook. DOI: 10.1201/9781003178903-15

Adetunji, Charles Oluwaseun, Wilson Nwankwo, Akinola Samson Olayinka, Olaniyan Tope Olugbemi, Muhammad Akram, Umme Laila, Michael SamuelOlugbenga, Ayomide Michael Oshinjo, Juliana BunmiAdetunji, Gloria E. Okotie, Nwadiuto (Diuto) Esiobu. (2022i). Machine Learning and Behaviour Modification for COVID-19. In: *Medical Biotechnology Biopharmaceutics, Forensic Science and Bioinformatics*. CRC Press. Pages 17. eBook. DOI: 10.1201/9781003178903-17

Akram, Muhammad, Ejaz Mohiuddin Charles Oluwaseun Adetunji, Tolulope Olawumi Oladosun, Phebean Ozolua, Frances Ngozi Olisaka, Chukwuebuka Egbuna, Olugbenga Michael, Juliana Bunmi Adetunji, Leena Hameed, Chinaza Godswill Awuchi, Kingsley Patrick-Iwuanyanwu, and Olugbemi Olaniyan. (2021a). Prospects of Phytochemicals for the Treatment of Helminthiasis. In: *Neglected Tropical Diseases and Phytochemicals in Drug Discovery*. Ch. 7. Wiley. DOI: 10.1002/9781119617143.ch7

Akram M, Adetunji CO, Egbuna C, Jabeen S, Olaniyan O, Ezeofor NJ, Anani OA, Laila U, Găman MA, Patrick-Iwuanyanwu K, Ifemeje JC, Chikwendu CJ, Michael OC, Rudrapal M. (2021b). Dengue Fever. In *Neglected Tropical Diseases and Phytochemicals in Drug Discovery*. Ch17. DOI: 10.1002/9781119617143

Arowora KA, Abiodun AA, Adetunji CO, Sanu FT, Afolayan SS, Ogundele BA (2012). Levels of aflatoxins in some agricultural commodities sold at Baboko Market in Ilorin, Nigeria. *Global Journal of Science Frontier Research*, 12(10):31–33.

Barbri NE, Mirhisse J, Ionescu R et al. (2009). An electronic nose system based on a micro-machined gas sensor array to assess the freshness of sardines. *Sensors and Actuators B*, 141(2): 538–543.

Benedetti S, Iametti S, Bonomi F, and Mannino S. (2005). Head space sensor array for the detection of aflatoxin M1 in raw ewe's milk, *Journal of Food Protection*, 68(5): 1089–1092.

Bibri SE, Krogstie J. 2020. Environmentally data-driven smart sustainable cities: Applied innovative solutions for energy efficiency, pollution reduction, and urban metabolism. *Energy Inform* 3: 29. https://doi.org/10.1186/s42162-020-00130-8.

Bucur B, Munteanu FD, Marty JL, and Vasilescu A. (2018). Advances in Enzyme-Based Biosensors for Pesticide Detection . *Biosensors (Basel)*, 8(2): 27. doi:10.3390/bios8020027

Chauhan N, Hooda V, and Pundir CS. (2012). Influence of copper nanoparticles on copper requiring enzyme. *Journal of Experimental Nanoscience*, 7: 6.

Chauhan N, Narang J, and Jain U. (2015). Amperometric acetylcholinesterase biosensor for pesticides monitoring utilising iron oxide nanoparticles and poly(indole-5-carboxylic acid. 111–122. https://doi.org/10.1080/17458080.2015.1030712.

Cheli F, Campagnoli A, Pinotti L, Savoini G, and Dell'Orto V. (2009). Electronic nose for determination of aflatoxins in maize. *Biotechnology, Agronomy and Society and Environment*, 13: 39–43.

Concina I, Bornšek M, Baccelliere S, Falasconi M, Gobbi E, and Sberveglieri G. (2010). Detection in soft drinks by Electronic Nose. *Food Research International*, 43(8): 2108–2114.

Concina, I., M. Falasconi, E. Gobbi et al. (2009). Early detection of microbial contamination in processed tomatoes by electronic nose. *Food Control*, 20(10): 873–880.

Dauda WP, Morumda D, Abraham P, Adetunji CO, Ghazanfar S, Glen E, Abraham SE, Peter GW, Ogra IO, Ifeanyi UJ, Musa H, Azameti MK, Paray BA, Gulnaz A. (2022a). Genome-Wide Analysis of Cytochrome P450s of Alternaria Species: Evolutionary Origin, Family Expansion and Putative Functions. *Journal of Fungi*, 8(4):324. https://doi.org/10.3390/jof8040324

Dauda WP, Abraham P, Glen E, Adetunji CO, Ghazanfar S, Ali S, Al-Zahrani M, Azameti MK, Alao SEL, Zarafi AB, Abraham MP, Musa H. (2022b) Robust Profiling of Cytochrome P450s (P450ome) in *Notable Aspergillus spp. Life*, 12(3):451. https://doi.org/10.3390/life12030451

EPA. (2016). Environmental Protection Agency. *What are pesticides and how do they work?* www.epa.nsw.gov.au/pesticides/pestwhatrhow.htm#pestdefns.

Evans P, Persaud KC, McNeish AS, Sneath RW, Hobson N, and Magan N. (2000). Evaluation of a radial basis function neural network for the determination of wheat quality from electronic nose data. *Sensors and Actuators B*, 69(3): 348–358.

Falasconi, M., E. Gobbi, M. Pardo, M. Della Torre, A. Bresciani, and G. Sberveglieri. (2005). Detection of toxigenic strains of *Fusarium verticillioides* in corn by electronic olfactory system, *Sensors and Actuators B*, 108(1–2): 250–257.

Gill H and Garg H (2013). Pesticides: Environmental Impacts and Management Strategies. DOI: 10.5772/57399

Gobbi E, Falasconi M, Torelli E, and Sberveglieri G. (2011). Electronic nose predicts high and low fumonisin contamination in maize cultures, *Food Research International*, 44: 992–999.

Grattieri M. and Minteer SD (2006). Self-Powered Biosensors. *FEMS Microbiology Reviews*, 30(3): 428–471, https://doi.org/10.1111/j.1574-6976.2006.00018.xACS Sens., 3, 44–53.

Grung M, Lin Y, Zhang H, Steen A, and Larssen T. (2015). Pesticide levels and environmental risk in aquatic environments in China – A review. *Environ Int*; 81:87–97. DOI: 10.1016/j.envint.2015.04.013

Hameed SW, Tahir MA, Kiran S, Ajmal S, Munawar A. (2016). Sensing and Degradation of Chlorpyrifos by using Environmental Friendly Nano Materials. *J BiosensBioelectron* 7:198. doi:10.4172/2155-6210.1000198

Hondred B, John A, Joyce B, Alves N, Scott AT, Walper SA, Medintz IL, and Jonathan C. (2018). Printed Graphene Electrochemical Biosensors Fabricated by Inkjet Maskless Lithography for Rapid and Sensitive Detection of Organophosphates. *ACS Applied Materials and Interfaces*. 10(13): 11125–11134. DOI: 10.1021/acsami.7b19763

Isaac OA, Adetunji CO. (2018). Production and evaluation of biodegraded feather meal using immobilised and crude enzyme from Bacillus subtilis on broiler chickens. *Brazilian Journal of Biological Sciences*, 5(10): 405–416.

Jeyabalan S, Hospet R, Thangadurai D, Adetunji CO, Islam S, Pujari N, Said Al-Tawaha ARM. (2021). Nanopesticides, Nanoherbicides, and Nanofertilizers: The Greener Aspects of Agrochemical Synthesis Using Nanotools and Nanoprocesses Toward Sustainable Agriculture. In: Kharissova, OV, Torres-Martínez LM, Kharisov BI. (eds) *Handbook of Nanomaterials and Nanocomposites for Energy and Environmental Applications*. Springer, Cham. https://doi.org/10.1007/978-3-030-36268-3_44

Jonsson A, Winquist F, Schnurer J, Sundgren H, and Lundstrom I. (1997). Electronic nose for microbial quality classification of grains, *International Journal of Food Microbiology*, 35(2): 187–193.

Ju H and Kandimalla V. (2008). Biosensors for Pesticides. In *Electrochemical Sensors, Biosensors and their Biomedical Applications*. Doi: 10.1016/B978-012373738-0.50004-0

Kaur J, Singh KV, Boro R, Thampi K, and Raje M. (2007). Immunochromatographic dipstick assay format using gold nanoparticles labeled protein-hapten conjugate for the detection of atrazine. *Environmental Science & Technology* 41: 5028–5036.

King AM, Aaron CK (2015). Organophosphate and carbamate poisoning. *Emergency Medicine Clinics of North America*, 33:133–151. doi: 10.1016/j.emc.2014.09.010

Kingsley EU, Adetunji CO, Bobadoye AO, Aigbe UO, Onyancha RB, Siloko IU, Emegha JO, Okocha GO, and Abiodun IC (2021). Bionanomaterials for biosensor technology. *Bionanomaterials. Fundamentals and biomedical applications*, pp. 5–22.

Kumar J, Mishra A, and Melo JS (2018). Biodegradation of Methyl Parathion and its Application in Biosensors. *Austin Journal of Environmental Toxicology*, 32(12): 234–256.

Labreche S, Bazzo S, Cade S, and Chanie E. (2005). Shelf life determination by electronic nose: Application to milk. *Sensors and Actuators B*, 106(1): 199–206.

Li C, Schmidt NE, and Gitaitis R. (2011). Detection of onion postharvest diseases by analyses of headspace volatiles using a gas sensor array and GC-MS, *Food Science and Technology*, 44(4): 1019–1025.

Marín S, Vinaixa M, Brezmes J, Llobet E, Vilanova X, Correig X, Ramos AJ, Sanchis V. (2007). Use of a MS-electronic nose for prediction of early fungal spoilage of bakery products. *International Journal of Food Microbiology*, 114(1): 10–16.

Muenchen D, Martinazzo J, Cezaro A, Rigo A. (2016). Pesticide Detection in Soil Using Biosensors and Nanobiosensors, Biointerphase. *Research in Applied Chemistry*, 6(6): 1659–1675. ISSN 2069–5837.

Nwankwo W, Adetunji CO, Ukhurebor KE, Panpatte DG, Makinde AS, Hefft DI. (2021). Recent Advances in Application of Microbial Enzymes for Biodegradation of Waste and

Hazardous Waste Material. In: Adetunji CO, Panpatte DG, Jhala YK (eds) *Microbial Rejuvenation of Polluted Environment. Microorganisms for Sustainability*, vol 27. Springer, Singapore. https://doi.org/10.1007/978-981-15-7459-7_3

Obare S, Chandrima D, Wen G, Haywood T, Samuels TA, Adams CP, Masika NO and Kenneth F. (2010). Fluorescent Chemosensors for Toxic Organophosphorus. *Pesticides: A Review Sensors*, 10(7): 7018–7043; https://doi.org/10.3390/s100707018

Okeke NE, Adetunji CO, Nwankwo W, Ukhurebor KE, Makinde AS, Panpatte DG (2021). A Critical Review of Microbial Transport in Effluent Waste and Sewage Sludge Treatment. In: Adetunji CO, Panpatte DG, Jhala YK (eds) *Microbial Rejuvenation of Polluted Environment. Microorganisms for Sustainability*, vol 27. Springer, Singapore. https://doi.org/10.1007/978-981-15-7459-7_10

Olaniyan OT, Adetunji CO, Adeniyi MJ, Hefft DI. (2022a). Machine Learning Techniques for High-Performance Computing for IoT Applications in Healthcare. In: *Deep Learning, Machine Learning and IoT in Biomedical and Health Informatics*. DOI: 10.1201/9780367548445-20. CRC Press. Pages 13. eBook.

Olaniyan OT, Adetunji CO, Adeniyi MJ, Hefft DI. (2022b). Computational Intelligence in IoT Healthcare. 2022. In: *Deep Learning, Machine Learning and IoT in Biomedical and Health Informatics*. DOI: 10.1201/9780367548445-19. CRC Press. Pages 13. eBook.

Olsson J, Börjesson T, Lundstedt T, and Schnürer J. 2000. Volatiles for mycological quality grading of barley grains: Determinations using gas chromatography-mass spectrometry and electronic nose. *International Journal of Food Microbiology*, 59(3): 167–178.

Oluwaseun AC, Phazang P, and Sarin NB (2017). Significance of Rhamnolipids as a Biological Control Agent in the Management of Crops/Plant Pathogens. *Current Trends in Biomedical Engineering & Biosciences*, 10(3): 54–55.

Oyedara OO, Adeyemi FM, Adetunji CO, Elufisan TO. (2022). Repositioning Antiviral Drugs as a Rapid and Cost-Effective Approach to Discover Treatment against SARS-CoV-2 Infection. In: *Medical Biotechnology, Biopharmaceutics, Forensic Science and Bioinformatics*. CRC Press. Pages 12. eBook. DOI: 10.1201/9781003178903-10.

Paolesse R, Alimelli A, Martinelli E. et al. (2006). Detection of fungal contamination of cereal grain samples by an electronic nose. *Sensors and Actuators B*, 119(2): 425–430.

Poirier L, Brun L, Jacquet P. (2017). Enzymatic degradation of organophosphorus insecticides decreases toxicity in planarians and enhances survival. *Sci Rep* 7: 15194 https://doi.org/10.1038/s41598-017-15209-8

Sassolas A, Prieto-Simón B, Marty JL. (2012). Biosensors for pesticide detection: new trends to pesticide. *American Journal of Analytical Chemistry*, 3: 210–232. http://dx.doi.org/10.4236/ajac.2012.33030

Mavrikou S, Flampouri K, Moschopoulou G, Mangana O, Michaelides A and Kintzios S. (2008). Assessment of organophosphate and carbamate pesticide residues in cigarette tobacco. *Novel Cell Biosensor Sensors*, 8(4): 2818–2832. https://doi.org/10.3390/s8042818

Stephenson GR, Ferris IG, Holland PT, Nordberg M. (2006). Glossary of terms relating to pesticides, *Pure and Applied Chemistry*, 78(11): 2075–2154.

Stoytcheva M. (2010). Enzyme vs. bacterial electrochemical sensors for organophosphorus. *Pesticides Quantification*. DOI: 10.5772/7155

Ukhurebor KE, Mishra P, Mishra RR, Adetunji CO (2020). Nexus between climate change and food innovation technology: recent advances. In: Mishra P, Mishra RR, Adetunji CO. (eds) *Innovations in Food Technology*. Springer, Singapore. https://doi.org/10.1007/978-981-15-6121-4_20

Ukhurebor KE, Nwankwo W, Adetunji CO, Makinde AS (2021). Artificial intelligence and internet of things in instrumentation and control in waste biodegradation plants: recent developments. In: Adetunji CO, Panpatte DG, Jhala YK (eds) *Microbial Rejuvenation of Polluted Environment. Microorganisms for Sustainability*, vol 27. Springer, Singapore. https://doi.org/10.1007/978-981-15-7459-7_12

Zamora-Sequeira R, Starbird-Perez R, Rojas-Carrillo O, and Vargas S. (2019). Molecules 24, 14: 2659. What are the main sensor methods for quantifying pesticides in agricultural activities? *A Review*. doi: 10.3390/molecules24142659

8 Artificial Intelligence in Food Fraud and Traceability

Ogundolie Frank Abimbola, Michael O. Okpara, Manjia Jacqueline Njikam, and Aruwa Christiana Elejo

CONTENTS

8.1 INTRODUCTION

Food fraud (FFD) and traceability are inherently linked in the sense that, fraud is the problem, while traceability is the way to resolving the problem (Zheng et al., 2021). Several definitions exist for food fraud; they share common views and can be broadly described as any deceitful dilution or addition to a raw material or food product, fraudulent manipulation, intentional or unintentional substitution, or misrepresentation of the food product for monetary gain (this can be achieved by either inflating the apparent value or decreasing production cost of the product) (González-Pereira et al., 2021).

Similarly, according to the description of Codex Alimentarius, food traceability is "the ability to follow the movement of a food through specified stage(s) of production, processing and distribution", while European Commission stated that 'traceability' is the "ability to trace and follow a food, feed, food-producing animal or substance intended to be, or expected to be incorporated into a food or feed, through all stages of production, processing & distribution" (Regulation (EC) No 178/2002, EC 2002).

DOI: 10.1201/9781003207955-8

Traceability can be carried out during supplies, processing, and consumption levels as a measure to uncover food fraud, or to prevent food fraud, in which case, it goes a long way to protect brands (Corina, 2013).

With the steady rise in the world population, it has been projected that there will be a 98 percent rise in the demand for food by 2050 (Mavani et al., 2021), and this is expected to lead to a greater extent, a more complex production chain that is insensitive to safety concerns such as contamination of foods, disease spread such as pandemics or epidemics in human beings, aquatic lives, domesticated and non-domesticated animals, deceitful farming methods, food fraud – even food authenticity-related problems (Mavani et al., 2022).

When dealing with the control of epidemics, some factors such as monitoring, traceability and data analysis to ensure food safety is of great importance. These factors also are considered when preventing an outbreak/epidemic (Zheng et al., 2021).

8.2 FOOD FRAUD AND TRACEABILITY

To achieve food security in global society today, efforts need to be put in place to ensure proper control of the food system. The common practice in the food sector is the use of misleading information provided on the packaging materials of several packaged foods. This is also associated with food security or food harm (Lord et al., 2017). This is of more concern because the rate of consumption of packaged foods has greatly increased over the years. Today, occurrences of fortified food content or packaging of adulterated items are now on the rise.

Thakur et al. (2017) observed a case of food fraud when they analyzed the contents of canned food with the expectation that it would contain deer meat; upon analysis, they discovered that the meat in the can was from a pig. This kind of food fraud can be harmful to humans, especially for those that have serious allergies to some specific meals/compounds.

Food fraud has been attributed to several packaged products, which includes wine, beverages, meat, poultry, dairy and fish products (Roberts, 1994). Over the decades, the main purpose of food fraud is in the area of commercial crime. As explained by both it entails turning a blind eye to the health danger of the food package and concentrating solely on maximizing the profit from the foods to ensure wealth generation (Naylor, 2003; Lord et al., 2017). Alteration of results, misinformation, production of counterfeit/substandard goods, adulteration, forging importation documents, or diversion are the common methods of food fraud used in various food processing industries (Spink, J., and Moyer, 2011; Obbink et al., 2014; Lord et al., 2017).

Despite the numerous and emerging forms of FFD occurring over the years, it can generally be categorized into two major types (Obbink et al., 2014). The type that involves repackaging of old and expired products, which are harmful to the end user (consumers), and the intentional misrepresentation of packed foods caused by adulteration, diversion of the original foods, production of substandard goods, or use of misleading packaging information, either by providing for content or by labels misrepresenting the contents.

8.2.1 Repackaging

This type of food fraud is the most dangerous because expired foods are being repackaged and rebranded for sale (Obbink et al., 2014). In many cases, various diseases are associated with the consumption of such expired food products.

In drugs, the repackaging of expired products, or products without active ingredients Williams and McKnight, 2014), is fast becoming of growing concern, seeking different regulations and working effortlessly on ensuring a system for easy detection of such products. That system is readily available to the consumers (Kaoud and Kurdi, 2022). A common method used in addressing this is in the application of barcodes on the packages, which can be used to trace the production line, quality control check, information about the products, and finally confirm the genuineness of the product. In Nigeria, some specific drugs that are highly prone to food fraud are packaged with barcodes that can be unveiled by the end-user, and the serial number can be used to ascertain the authenticity of the product.

8.3 FOOD TRACEABILITY

In recent years, due to the several incidences of FFD and safety issues, the use of food traceability has recently evolved as a key mechanism for monitoring food products across the food supply chain. In the food sectors, delivery of food products from the manufacturer or farmer to the consumer requires strong technical expertise, robust database management, reliable data, and improved food safety culture. Perhaps, the best way to address the problems associated with food fraud and safety issues is to establish a solid and reliable food traceability system. The traceability system ensures that high food quality and safety are guaranteed, and this could solve the problem of trust between consumers and manufacturers. Studies relating to food traceability systems today are dependent on availability of big data using artificial intelligence (AI) and the Internet of Things (IoT) which identifies issues and suggests ways to address these challenges. Some of which are posed by data storage or inadequate information/statistics in the local or conventional traceability systems and also low credibility of these conventional systems (Rejeb et al., 2022).

Over the past decades, AI has been suggested as a viable technology to advance and transform food safety practices. The application of AI in various tasks such as food quality determination, control tools, classification of food, and prediction purposes has intensified their demand in the food industry. Artificial intelligence involves the capacity of computers to emulate the human learning process, knowledge, or data storage for smart execution of commands. Artificial intelligence systems utilize data provided to analyze a given condition or environment then effectively take actions. These systems are designed to be independent, as they have some level of freedom to carry out set goals. They use big data to improve their behavior and increase their problem-solving abilities, which includes effective execution of targeted goals/tasks (Mavani et al., 2022).

AI-based systems can either be software-based or embedded in hardware devices. AI-based systems have a wide range of algorithms to choose from such as expert systems, reinforcement learning, and fuzzy logic (FL), Turing test, swarm intelligence, cognitive science, artificial neural network (ANN), and logic programming. The effective and efficient performance of AI has made it the most favorable tool to use in industries for decision making and process estimation aimed at overall cost reduction, quality enhancement, and improved profitability.

8.4 IMPORTANCE OF FOOD TRACEABILITY

Food fraud is a menace and deserves more attention due to the numerous health and economic risks customers face because of this problem. In most cases, and depending on the kind of food fraud committed, severe illness, food allergies, and even death can be caused by food fraud. On the other hand, food traceability of foods from farm to fork is important when considering that the food products supplied to the public have to meet basic food regulatory demands, for consumer protection and safety (Chhikara et al., 2018).

Additionally, as a vital tool for each food business, food traceability provides information within the business for process control and management. More importantly, food traceability enhances stock control, efficient translocation of food products, effective quality control, collection, collation, and storage of information about food ingredients and products. Food traceability is an essential component in the food manufacturing industry which ensures the safety of domestic and global food supply and assures the customers of brand protection. Every food manufacturer must, therefore, ensure that their food traceability systems are up to the highest standards to protect the brand image in the market while also guaranteeing food safety.

Consequently, these growing requirements are pushing food manufacturers and processors to maintain upstream and downstream traceability in the food supply chain.

Food companies, the primary challenge(s) faced daily, revolves around the development of more cost-effective and safe methods of ensuring optima traceability of their products. Depending on the products being transported or manufactured, some companies utilize traceability systems to ensure raw materials are properly traced along the line of production down to the consumers. Traceability can enable experts or administrators to quickly differentiate dangerous goods and also ascertain their level of deterioration.

Food traceability must be treated as a very important component in the food supply chain to minimize risk factors to the producers and consumers. Food traceability will help the manufacturers/suppliers track the food product and also increase consumer acceptability (Chhikara et al., 2018).

8.5 APPLICATION OF ARTIFICIAL INTELLIGENCE IN FOOD TRACEABILITY

The use of AI in food traceability can be applied for various purposes in the different food industries (Mavani et al., 2021). AI-assisted food traceability can be applied between the inventory and utilization of starting materials, the received unit, usually referred to as the raw material unit, and the final product. It can be used to trace the source of the units, the supplier, the buyer of the units sold; or even applied throughout the food chain (Mavani et al., 2022). Within the past decade, there has been a growing global interest in AI and business automation because of increased access to and generation of data. In like manner, advances in computing power have sparked new research, investment, and applications across many industrial and service sectors, including the food industry (Sharma et al., 2021).

Application of artificial intelligence in the food sector ensures that enormous improvement is achieved during food production and processing. With automation, food manufacturers are rapidly growing their businesses by predicting market situations while concentrating their efforts heavily on logistics, which largely includes the various supply chains, has resulted in a shift in the consumer consumption patterns, especially for the products with short shelf lives (Mavani et al., 2022). Although the implementation of the use of modern technologies such as AI, machine learning and big data is far from been fully executed in these industries, companies using this technology have been reporting a better profit margin compared to those using the conventional techniques and/or methods.

8.6 APPLICATION OF AI IN FOOD FRAUD; IMPLICATIONS OF FOOD FRAUD

Food fraud is an old menace whose origins are very difficult to trace. However, over the centuries and decades, the incidence of food fraud has continued to evolve, especially as policymakers in governments come up with new regulations to checkmate the menace. Despite being under-reported because of the technicality involved in identifying food fraud, the effects of food fraud are always far-reaching with economic, health, and social and religious implications. Food fraud occurs in many food products including fruits, vegetables, meats, fish, spices and chili powder, alcoholic drinks, fruit juice, baby foods, beverages, soft drinks, rice, canned foods, and so on. Over the years, there have been many reported cases of food fraud that have led to deaths, hospitalizations, huge financial losses, termination of business contracts, recall of products from the markets, customer distrust and so on. Herein the economic, health, social and religious implications of food fraud are summarized with a few examples that were reported.

8.7 HEALTH IMPLICATION

Perhaps, one of the biggest outcomes of food fraud all over the world is the huge impact it has on public health. Over the years, public health concerns regarding the risk associated to food fraud has been on the rise. There have been several reported cases of food fraud that have had different health implications. A notable example is the European horsemeat food fraud of 2013, where beef was adulterated with horsemeat and other adulterants. Further investigations revealed that some batches of the horsemeat already on the market contained phenylbutazone (BBC, 2013b). Phenylbutazone is a cyclooxygenase inhibitor, and also a non-steroidal anti-inflammatory drug (NSAID) commonly administered to animals for treating short-term pain and fever. However, the use of phenylbutazone for humans was severely restricted in the United States due to reported concerns that the compound could be an inducer of some blood dyscrasias such as leukopenia, thrombocytopenia, aplastic anaemia, and agranulocytosis. In some cases, the use of phenylbutazone led to some fatalities (Lees and Toutain, 2013). A study by Kari et al. (1995) also revealed that phenylbutazone is a potential carcinogen that was linked to the onset of tumours in rats and mice kidneys and livers, respectively (Kari et al., 1995).

Another incidence of food fraud with massive public health implications occurred in China in 2008 when powdered infant milk was deliberately adulterated with cyanuric acid and scrap-grade melamine. Cyanuric acid is a member of the triazine family of compounds and is produced as an intermediate during the synthesis or degradation of melamine. Melamine is a nitrogen-rich compound that is used to manufacture several plastic products, coatings, adhesives, kitchenware, paints, and fertilizer mixtures. Cyanuric acid and melamine were added to dairy products to falsely exaggerate the amount of protein present in the various dairy products. This act of food fraud was responsible for the hospitalization of 52,000 children out of about 300,000 children who fell ill with 6 deaths recorded and several cases of acute renal failure or urinary tract stones confirmed among the hospitalized children (BBC, 2010; Gossner et al., 2009).

Although there is still no study that has investigated the oral toxicity of melamine in humans, studies in rats have reported a lethal dose, LD_{50} of 0.316g/kg of body weight. However, investigations into the amount of melamine added to the infant milk product in 2008 revealed that melamine concentration ranging from 0.15 to 4.7 g/kg with a median of 1.9 g/kg detected in the milk products tested (WHO, 2008). A major contributor to the health hazard of melamine in the milk product was the formation of a toxic cyanuric acid-melamine co-crystalline complex through hydrogen bonds. The cyanuric acid-melamine co-crystalline complex formed a precipitate that accumulated in the renal tubules and caused renal failure in many hospitalized children (Dobson et al., 2008; Bhalla et al., 2009).

A common and conventional way of artificially enhancing the ripening of some fruits involves the application of calcium carbide. This has been a common practice in many developing countries. This act of food fraud is mostly committed for quick sales of food products owing to the increased demand for fruits and vegetables. The commercial production of calcium carbide was intended for the desulphurization stage in steel production, welding, and the synthesis of calcium cyanamide and acetylene. But its use in fruit and vegetable ripening is an act of food fraud with health implications that can be deleterious.

During the process of utilizing calcium carbide for artificial and quick ripening of some vegetables or fruits, the dissolution of calcium carbide in water produces acetylene. Acetylene is an analogue of ethylene, which is a natural fruit-ripening agent produced by fruits. In the presence of oxygen, the acetylene acts as a sedative and could lead to hypoxia, headache, memory loss, loss of consciousness, cerebral oedema, and lactic acidosis (Per et al., 2007). Commercially produced calcium carbide has been reported to contain heavy metals like arsenic and traces of phosphorus hydride (Per et al., 2007; Bandyopadhyay et al., 2013; Maduwanthi and Marapana, 2019). Traces of arsenic could be stored in the fruits and vegetables subjected to calcium carbide treatment. Consumption of such contaminated fruits or vegetables could lead to the accumulation of arsenic in the muscles, kidneys, heart, lungs, and liver. Over time, the bioaccumulation of non-metabolized arsenic in the body marks the onset of arsenic neurotoxicity, hepatotoxicity, nephrotoxicity, and carcinogenicity (Okeke et al., 2022).

One form of food fraud that is mostly unreported in many parts of the world is committed by food vendors. In their quest to maximize profit, many food vendors intentionally purchase substandard or expired or rotten food ingredients because the ingredients are sold at a ridiculously cheaper rate. Rotten tomatoes, pepper, onions, cabbage, meats, fish, eggs, and expired canned ingredients like green peas, spices, and food additives are examples of food ingredients that fraudulent food vendors purchase to cut costs. Consumption of food prepared with substandard/expired/rotten food ingredients can lead to food poisoning or foodborne illnesses because they contain harmful microbes and chemical substances.

Microbes like *Vibrio cholerae*, Enterohaemorrhagic *Escherichia coli*, *Salmonella*, *Shigella*, *Campylobacter*, *V. parahaemolyticus*, *Staphylococcus aureus*, *Listeria*, norovirus, trematodes, nematodes, *Entamoeba histolytica*, *C. perfringens*, *Echinococcus* spp., and *Clostridium botulinum*; chemicals like mycotoxins and biotoxins and heavy metals like arsenic, mercury, and lead. are examples of food contaminants associated with foodborne illnesses. According to a report by the World Health Organization (WHO), at least 600 million people in the world get sick from foodborne illnesses with about 420, 000 deaths recorded annually (WHO, 2020). Common foodborne illnesses include *Salmonellosis*, *shigellosis*, *campylobacteriosis*, *listeriosis*, giardiasis, vibrio infection, norovirus infection, viral gastroenteritis, and so on, and some common symptoms associated with foodborne illnesses are diarrhea, nausea, weight loss, bloating, dehydration, fatigue, headache, loss of appetite, and abdominal cramps.

In some West African countries, particularly Nigeria and Ghana, systematic adulterated of crude palm oil by their manufacturers has been reported using potash, red dye and Sudan IV dye with the sole aim of maximizing profits (Andoh et al., 2019; Kola-Ajibade et al., 2021; MacArthur et al., 2021). Though Nigeria and Ghana produce one of the best natural palm oils in the world, the high demand for palm oil from these countries contributed to the fraudulent acts of adulterating crude palm oil from Nigeria and Ghana with Sudan IV dye. Sudan IV dye, a lysochrome diazo dye that is widely utilized as a color pigment for the commercial production of soaps, shoe and floor polishes, inks, plastics, and textiles among others.

The addition of Sudan IV dye in crude palm oil gives an exaggerated red coloration that deceives the unsuspecting retailer or consumer into believing that the oil is 100 percent natural palm oil. The health implication of consuming Sudan IV can be fatal over time. The consumption of Sudan IV dye has been reported to cause impairment of hepato-renal functions, over-expression of pro-inflammatory cytokines, induction of reactive oxygen species, and inhibition of the activity of antioxidant enzyme catalase (Eteng et al., 2022; Kola-Ajibade et al., 2021; Li et al., 2017).

Adulteration of alcoholic drinks with methanol is a common food fraud activity that is practiced in many countries. Methanol is a harmful alcohol when accumulated in large quantities; it is used as a renewable energy source for fueling automobiles and ships, for plastic production, and also in making pesticides (Dalena et al., 2018; Simon Araya et al., 2020). Methanol is added to wines and other alcoholic drinks by food fraudsters because it is relatively cheaper and highly intoxicating. However, the health implication of methanol poisoning can be fatal. Upon ingestion, methanol can be oxidized to formaldehyde by the enzymatic activity of alcohol oxidase and catalase. And formic acid production through the action of aldehyde dehydrogenase on formaldehyde (Li et al., 2021).

Formation of formic acid as a metabolite is the basis for the toxicity of methanol. Methanol poisoning has been linked to loss of vision and restricted diffusion of the optic nerves, metabolic acidosis, multiple organ failures, damage to the central nervous system and nerve cell degeneration, intracranial hemorrhage, coma, and death (Chan and Chan, 2018; Mojica et al., 2020; Rahimi et al., 2021). Over the years, there have been several reported cases of methanol poisoning that led to the death of scores of consumers in different countries (Hassanian-Moghaddam et al., 2015; Rahimi et al., 2021; Rostrup et al., 2016; Zakharov et al., 2014).

8.8 ECONOMIC IMPLICATION

The economic implication of food fraud cannot be overlooked. Food manufacturers are faced with different kinds of economic losses owing to food fraud. These can be from loss of income revenue due to the circulation of adulterated/counterfeit and relatively cheaper alternatives in the market by food fraudsters, or the recall of substandard food products, or the outright cancellation of business contracts. According to the Grocery Manufacturers' Association, one out of every ten food products available on the market is affected by food fraud. And these products of food fraud are reported to cost at least $10 billion per annum (Johnson, 2014; Manning, 2016).

An example of a food fraud case with huge economic implications was the European horsemeat food fraud of 2013, which involved many meat-processing/supplying companies in the UK, France, Belgium, Italy, Spain, Switzerland, Luxembourg, Latvia, Lithuania, and The Netherlands. In this food-fraud case, horsemeat, pork, and other adulterants were reportedly packaged and labelled as beef, as revealed by the detection of equine and porcine DNA in some of the meat samples tested (BBC, 2013a; FSAI, 2013). The horsemeat food fraud case led to the recall and destruction of the adulterated meat from the market with enormous economic losses recorded by the companies. During this period, the sales of beef were negatively affected as the consumers were skeptical about meat products labelled as beef. Thus, causing low revenues for authentic beef processing companies and suppliers across Europe (BBC, 2013b; Regan et al., 2015).

In many instances of food fraud involving a manufacturer or even a retailer, some consumers assume that every other food product supplied or sold by the manufacturer and/or retailer could also be a product of food fraud. This can eventually lead to a reduction in sales and possible expiration of other products on offer from the manufacturer and/or retailer even when the other products are authentic. The economic impact is usually worse on the manufacturer when there are alternative products that the customers can purchase in the market.

For instance, during the melamine milk scandal in China, the sales of locally produced milk reduced by almost 40 percent while the sales of imported milk increased from 37 to 47 percent suggesting that some consumers developed a preference and trust for imported milk over the locally produced milk owing to the food fraud incidence that involved locally produced milk (Qiana, et al., 2011). Other economic implications of the Chinese melamine milk scandal included the Chinese government's seizure of more than two thousand tons of adulterated milk products while around nine thousand tons of the same product were recalled by a supplier (Gossner et al., 2009). The economic impact of this food fraud was far-reaching as dozens of countries either banned melamine-containing foods or recalled those already in circulation (Bhalla et al., 2009).

Another economic implication of food fraud is the possible termination of bilateral trade agreements between countries upon the detection of contamination/adulteration in imported food products. In 2004, the Sudan IV dye food fraud activity in Ghana was flagged by the European Commission's Rapid Alert System for Food and Feed (RASFF). In their 2004 report, RASFF notified the European Union of the rise in the adulteration of imported palm oil with Sudan IV dye. This prompted many European countries to ban the importation of palm oil from Ghana – the major African exporter of palm oil to member states of the European Union (RASFF, 2004). The ban on the importation of palm oil from Ghana had some economic consequences on Ghana as the foreign exchange derived from the export of palm oil was affected owing to food fraud activities.

8.9 SOCIAL IMPLICATIONS

Food fraud also has some social implications that determine how consumers perceive a particular food manufacturer involved in intentional or unintentional food fraud activity. Following the abovementioned European horsemeat food fraud of 2013, there was huge customer dissatisfaction and a loss of credence towards the beef suppliers. Also, in the European beef supply chain, the extensive media outcry over the incident contributed immensely to the consumer's distrust and made the public concerns about the integrity of those involved. (Manning, 2016).

8.10 RELIGIOUS IMPLICATIONS

Some religions have certain dietary restrictions that believers are expected to strictly obey. Some cases of food fraud where the fraudsters have intentionally, or unintentionally. adulterated or mislabeled food items have caused people of a particular religious faith to consume food items that are considered sacrilegious to their religious beliefs. Again, during the European horsemeat food fraud, when beef was adulterated with other undeclared meats like horsemeat and pork, unsuspecting Jews and Muslims who consumed the adulterated beef were made to break a law in Judaism and Islam that forbids the consumption of pork.

There are many cases of food fraud that go unreported, either because the consumers and suppliers do not possess the necessary tools for testing the quality of the food or because the government agencies/authorities responsible for ensuring high quality of food have been compromised by manufacturers of substandard/adulterated food. Whichever is the case, the consumers tend to suffer most from the effects of food fraud.

REFERENCES

Andoh, S. S., Nuutinen, T., Mingle, C., & Roussey, M. (2019). Qualitative analysis of Sudan IV in edible palm oil. *Journal of the European Optical Society*, 15(21), 1–5. https://doi.org/10.1186/s41476-019-0117-0

Bandyopadhyay, S., Saha, M., Biswas, S., Ranjan, A., Naskar, A. K., & Bandyopadhyay, L. (2013). Calcium carbide related ocular burn injuries during mango ripening season of West Bengal, Eastern India. *Nepalese Journal of Ophthalmology*, 5(10), 242–245. https://doi.org/10.3126/nepjoph.v5i2.8736

BBC. (2010). China dairy products found tainted with melamine. Retrieved March 14, 2022, from www.bbc.com/news/10565838

BBC. (2013a). Find beef lasagne contained up to 100% horsemeat, FSA says. Retrieved March 14, 2022, from www.bbc.com/news/uk-21375594

BBC. (2013b, April 10). Veterinary drug found in Asda budget corned beef. Retrieved March 14, 2022, from www.bbc.com/news/uk-22087123

Bhalla, V., Grimm, P. C., Chertow, G. M., & Pao, A. C. (2009). Melamine nephrotoxicity: An emerging epidemic in an era of globalization. *Kidney International*, 75(8), 774–779. https://doi.org/10.1038/ki.2009.16

Chan, A. P. L., & Chan, T. Y. K. (2018). Methanol as an unlisted ingredient in supposedly alcohol-based hand rub can pose a serious health risks. *International Journal of Environmental Research and Public Health*, 15(7), 6–11. https://doi.org/10.3390/ijerph15071440

Chhikara, N., Jaglan, S., Sindhu, N., Anshid, V., Charan, M. V. S., & Panghal, A. (2018). Importance of traceability in food supply chain for brand protection and food safety systems implementation. *Annals of Biology*, 34(2), 111–118.

Dalena, F., Senatore, A., Marino, A., Gordano, A., Basile, M., & Basile, A. (2018). Methanol production and applications: An overview. *Methanol*, 3–28.

Dobson, R. L. M., Motlagh, S., Quijano, M., Cambron, R. T., Baker, T. R., Pullen, A. M., … Daston, G. P. (2008). Identification and characterization of toxicity of contaminants in pet food leading to an outbreak of renal toxicity in cats and dogs. *Toxicological Sciences*, 106(1), 251–262. https://doi.org/10.1093/toxsci/kfn160

Ene, C. (2013). The relevance of traceability in the food chain. *Economics of Agriculture*, 60(2), 287–297.

Eteng, O. E., Moses, C. A., Ugwor, E. I., Enobong, J. E., Akamo, A. J., Adebekun, Y., … Ubana, E. (2022). Ingestion of Sudan IV-adulterated palm oil impairs hepato-renal functions and induces the overexpression of pro-inflammatory cytokines: A sub-acute murine model. *Egyptian Journal of Basic and Applied Sciences*, 9(1), 11–22. https://doi.org/10.1080/2314808X.2021.2010884

FSAI. (2013). FSAI survey finds horse DNA in some beef burger products. Retrieved March 14, 2022, from https://web.archive.org/web/20130119080947/; www.fsai.ie/news_centre/press_releases/horseDNA15012013.html

González-Pereira, A., Otero, P., Fraga-Corral, M., Garcia-Oliveira, P., Carpena, M., Prieto, M. A., & Simal-Gandara, J. (2021). State-of-the-art of analytical techniques to determine food fraud in olive oils. *Foods*, 10, 484. https://doi.org/10.3390/foods10030484

Gossner, C. M. E., Schlundt, J., Ben Embarek, P., Hird, S., Lo-Fo-Wong, D., Beltran, J. J. O., Teoh, K. N., & Tritscher, A. (2009). The melamine incident: Implications for international food and feed safety. *Environmental Health Perspectives*, 117(12), 1803–1808. https://doi.org/10.1289/ehp.0900949

Hassanian-Moghaddam, H., Nikfarjam, A., Mirafzal, A., Saberinia, A., Nasehi, A. A., Masoumi Asl, H., & Memaryan, N. (2015). Methanol mass poisoning in Iran: Role of case finding in outbreak management. *Journal of Public Health* (United Kingdom), 37(2), 354–359. https://doi.org/10.1093/pubmed/fdu038

Johnson, R. (2014). Food fraud and "Economically motivated adulteration" of food and food ingredients. *CRS Report.*

Kaoud, A. H., & Kurdi, F. T. (2022). Repackaging products to mislead the consumer. *Tikrit University Journal for Rights*, 6(2), 116–144.

Kari, F., Bucher, J., Haseman, J., Eustis, S., & Huff, J. (1995). Long-term exposure to phenylbutazone induces kidney tumours in rats and liver tumours in mice. *Japanese Journal of Cancer Research*, 86(3), 252–263. https://doi.org/10.1111/j.1349-7006.1995.tb03048.x

Kola-Ajibade Ibukun, R., Atere, G., & Olusola Augustine, O. (2021). Effects of azo dye adulterated palm oil on the expression of inflammatory, functional, antioxidant markers and body weights in albino rats. *Journal of Toxicology and Risk Assessment*, 7(1), 1–6. https://doi.org/10.23937/2572-4061.1510041

Lees, P., & Toutain, P.-L. (2013). Pharmacokinetics, pharmacodynamics, metabolism, toxicology and residues of phenylbutazone in humans and horses. *Veterinary Journal*, 196(3), 294–303. https://doi.org/10.1016/j.tvjl.2013.04.019

Li, T., Hao, M., Pan, J., Zong, W., & Liu, R. (2017). Comparison of the toxicity of the dyes Sudan II and Sudan IV to catalase. *Journal of Biochemical and Molecular Toxicology*, 31(10), 1–8. https://doi.org/10.1002/jbt.21943

Li, T., Wei, Y., Qu, M., Mou, L., Miao, J., Xi, M., Liu, Y., & He, R. (2021). Formaldehyde and de/methylation in age-related cognitive impairment. *Genes*, 12(6), 913.

Lord, N., Spencer, J., Albanese, J., & Flores Elizondo, C. (2017). In pursuit of food system integrity: The situational prevention of food fraud enterprise. *European Journal on Criminal Policy and Research*, 23(4), 483–501.

MacArthur, R., Teye, E., & Darkwa, S. (2021). Quality and safety evaluation of important parameters in palm oil from major cities in Ghana. *Scientific African*, 13, e00860. https://doi.org/10.1016/j.sciaf.2021.e00860

Maduwanthi, S. D. T., & Marapana, R. A. U. J. (2019). Induced ripening agents and their effect on fruit quality of banana. *International Journal of Food Science*, 2019, 1–9. https://doi.org/10.1155/2019/2520179

Manning, L. (2016). Food fraud: policy and food chain. *Current Opinion in Food Science*, 10(2), 16–21. https://doi.org/10.1016/j.cofs.2016.07.001

Mavani, N. R., Ali, J. M., Othman S., Hussain M. A., Hashim, H., & Rahman, N. A. (2022). Application of artificial intelligence in food industry: a guideline. *Food Engineering Reviews*, 14, 134–175. https://doi.org/10.1007/s12393-021-09290-z

Mojica, C. V., Pasol, E. A., Dizon, M. L., Kiat Jr, W. A., Lim, T. R. U., Dominguez, J. C., Valencia, V. V., & Tuaño, B. J. P. (2020). Chronic methanol toxicity through topical and inhalational routes presenting as vision loss and restricted diffusion of the optic nerves on MRI: A case report and literature review. *Eneurologicalsci*, 20, 100258.

Naylor, R. T. (2003). Towards a general theory of profit-driven crimes. *British Journal of Criminology*, 43, 81–101.

Obbink, N., Frissen, J. M., & Post, S. B. (2014). Official Control: J. Food Frauds. *Meat Inspection and Control in the Slaughterhouse*, 628–638. https://doi.org/10.1002/9781118525821.ch24j

Okeke, E. S., Okagu, I. U., Okoye, C. O., & Ezeorba, T. P. C. (2022). The use of calcium carbide in food and fruit ripening: Potential mechanisms of toxicity to humans and prospects. *Toxicology*, 468(10), 1–12. https://doi.org/10.1016/j.tox.2022.153112

Per, H., Kurtoğlu, S., Yağmur, F., Gümüş, H., Kumandaş, S., & Poyrazoğlu, M. H. (2007). Calcium carbide poisoning via food in childhood. *Journal of Emergency Medicine*, 32(2), 179–180. https://doi.org/10.1016/j.jemermed.2006.05.049

Qiana, G., Guoa, X., Guob, J., & Wub, J. (2011). China's dairy crisis: Impacts, causes and policy implications for the sustainable dairy industry. *International Journal of Sustainable Development and World Ecology*, 18(5), 434–441. https://doi.org/10.1080/13504509.2011.581710

Rahimi, R., Zainun, K. A., Noor, N. M., Kasim, N. A. M., Shahrir, N. F., Azman, N. A., … Room, N. H. M. (2021). Methanol poisoning in Klang Valley, Malaysia: Autopsy case series. *Forensic Science International: Reports*, 3(1), 1–6. https://doi.org/10.1016/j.fsir.2021.100170

RASFF. (2004). Rapid alert system for food and feed (RASFF) annual report on the functioning of the RASFF. Retrieved from https://ec.europa.eu/food/system/files/2016-10/rasff_annual_report_2004_en.pdf

Regan, Á., Marcu, A., Cha, L. C., Wall, P., Barnett, J., & McConnon, Á. (2015). Conceptualising responsibility in the aftermath of the horsemeat adulteration incident: An online study with Irish and UK consumers. *Health, Risk and Society*, 17(2), 149–167. https://doi.org/10.1080/13698575.2015.1030367

Rejeb, A., Rejeb, K., Abdollahi, A., Zailani, S., Iranmanesh, M., & Ghobakhloo, M. (2022). Digitalization in food supply chains: A bibliometric review and key-route main path analysis. *Sustainability*, 14, 83. https://doi.org/10.3390/su14010083

Roberts, D. C. E. (1994). Food authenticity. *British Food Journal*, 96(9), 33–35. https://doi.org/10.1108/00070709410072490

Rostrup, M., Edwards, J. K., Abukalish, M., Ezzabi, M., Some, D., Ritter, H., Menge, T., Abdelrahman, A., Rootwelt, R., Janssens, B., & Lind, K. (2016). The methanol poisoning outbreaks in Libya 2013 and Kenya 2014. *PloS one*, 11(3), e0152676.

Sharma, S., Gahlawat, V. K., Rahul, K., Mor, R. S., & Malik, M. (2021). Sustainable innovations in the food industry through artificial intelligence and Big Data analytics. *Logistics*, 5, 66. https://doi.org/ 10.3390/logistics5040066

Simon Araya, S., Liso, V., Cui, X., Li, N., Zhu, J., Sahlin, S. L., Jensen, S. H., Nielsen, M. P., & Kær, S. K. (2020). A review of the methanol economy: The fuel cell route. *Energies*, 13(3), 596.

Spink, J., & Moyer, D. C. (2011). Defining the public health threat of food fraud. *Journal of Food Science*, 76(9), R157–R163.

Thakur, M., Javed, R., Kumar, V. P., Shukla, M., Singh, N., Maheshwari, A., Mohan, N., Wu, D. D., & Zhang, Y. P. (2017). DNA forensics in combating food frauds: A study from China in identifying canned meat labelled as deer origin. *Current Science*, 112, 2449–2452.

WHO. (2008). *Melamine and cyanuric acid: Toxicity, preliminary risk assessment and guidance on levels in food*. World Health Organization (Vol. 25), Geneva. Retrieved from www.who.int/foodsafety/fs_management/Melamine.pdf

WHO. (2020). Food safety. Retrieved March 14, 2022, from www.who.int/news-room/fact-sheets/detail/food-safety#:~:text=An estimated 600 million – almost, healthy life years (DALYs

Williams, L., & McKnight, E. (2014). The real impact of counterfeit medications. *US Pharmacists*, 39(6), 44–46.

Zakharov, S., Pelclova, D., Urban, P., Navratil, T., Diblik, P., Kuthan, P., Hubacek, J.A., Miovsky, M., Klempir, J., Vaneckova, M., & Seidl, Z. (2014). Czech mass methanol outbreak 2012: Epidemiology, challenges and clinical features. *Clinical Toxicology*, 52(10), 1013–1024.

Zheng, M., Zhang, S., Zhang, Y., & Hu, B. (2021). Construct food safety traceability system for people's health under the internet of things and big data. *IEEE/Access*, 9, 70571–70583. Doi:10.1109/ACCESS.2021.3078536

9 Relevance of Biosensor in the Detection of Contaminants and Pollutants in Food, Agricultural, and Environmental Sector

Inobeme Abel, Charles Oluwaseun Adetunji, Akinola Samson Olayinka, and Olalekan Akinbo

CONTENTS

9.1 INTRODUCTION

Environmental sensing for the purpose of monitoring has gradually witnessed significant advancement over time. There has been rapid evolution in the area of environmental sensors, and this has gained several areas of applications. Sensors have been employed for the monitoring of environmental conditions, to show reliable trends, aid in the development of reports for the making of policy makers and also help in supporting policy goals. Environmental sensing is the collection of data for better comprehension of environmental phenomena. Environmental monitoring has various advantages, which include reduction of pollution, minimization effects

DOI: 10.1201/9781003207955-9

of human activities and reducing occupational diseases (Adetunji et al., 2020; Adetunji et al., 2020a,b,c; Oyedara et al., 2022; Adetunji et al., 2022a,b,c,d,e,f,g,h; Adetunji and Oyeyemi, 2022; Olaniyan et al., 2022). Sensors are currently being employed in various communities for various environmental issues. Some of the area of applications includes sensing of oceans, forests, and volcanoes. Sensors have also been applied in the monitoring of air pollution. Sensors have been employed in different cities for monitoring of the contents of dangerous atmospheric gases. This takes advantage of the ad hoc wireless connections, thereby improving their mobility for the detection readings in varying areas (Kendler and Fishbain, 2022). They have also been deployed in the detection of fires in the forest. A collection of various networks of the sensors node could be installed for the detection of fire in the forest. Most environmental devices are usually equipped with sensors for the measurement of humidity, temperature and various gases that are produced by fires in vegetation. Early detection is pertinent; the fire fighters are able to know when the fire commences and the rate of spreading. They have also been deployed in the detection of landslides (Jain et al., 2020). Various systems for the detection of landslides makes use of sensors for the detection of slight motion of soil and variation in various factors, which may occur prior to or during landslides. The data collected makes it possible for the prediction of the occurrence of the sliding prior to its actual happening. There are numerous areas of emphasis of environmental sensors, which include monitoring water pollution, monitoring of air, pollution in urban environments, industrial emissions and monitoring of hazardous waste materials (Mitchell et al., 2017). The rapid advancement in the area of sensing systems and sensors has remarkable impact in the area of environment and agriculture. The impacts of applications of sensors vary from fine scales to global levels of remote sensing surveys of the environment. In recent times the manufacturers of various environmental machineries have been integrating a rising number of improved sensor techniques for precision in the environmental devices such as the miniature photosynthesis device, carbon flux soil monitoring, reflectometry and salinity monitoring in soil. Various kinds of environmental sensors have been fabricated, including temperature sensors, air flow sensors, humidity sensors, vibration sensors, pressure sensors, water leak sensors, contact closure sensors, amongst others (Lin et al., 2018). Environmental sensors are capable of measuring various environmental conditions, which include humidity and temperature amongst others. The real time data collected from the sensors are used for monitoring various trends in environmental phenomena and efficient predictions. The advancement of the area of Internet of Things (IoT) as well as maker motion has gradually paved the way for the development of cheap electronic sensors for the transformation of global monitoring of the environment. Alternative approaches in environmental monitoring involving the use of novel innovative techniques are gradually gaining much popularity in the field of environmental science (Parthipan and Dhanasekaran, 2019).

This chapter discusses the environmental aspects of sensors. It highlights the various areas of applications of sensors in the environment. It also discusses some of the advantages and limitations in the environmental usage of sensors. Finally it presents some of the future trends on application of sensors in solving environmental issues.

9.2 CLASSIFICATION OF BIOSENSORS

Various criteria have been employed in the classification of biosensors. Some of the classifications have been based on the biorecognition principles, while others are in terms of the transduction signal as shown in Figure 9.1. Based on the transduction component, the biosensors have been grouped into optical, thermal, electrochemical, piezoelectric types. The electrochemical biosensors have been further classified into the potentiometric and the amperometrics. The most commonly employed are the electrochemical followed by the optical. There are also catalytic biosensors that work on same principle as the electrochemical. The optical transducers make use of properties such as Raman scattering, florescence, biochemiluminescence and absorption of light and the surface resonance, and they possess the advantage in that it is possible to directly measure the analyte without the introduction of labeled molecules (Ranveer et al., 2015).

On the basis of the biorecognition principle, the classification of biosensors is into DNA, whole cell, enzymatic, nonenzymatic receptor and immunochemical. The immunochemical are known to have the advantage of selectivity and sensitivity. It however has the disadvantages that there are challenges with the regeneration of the cross reactivity and immunosurface. Enzymes are commonly employed as the recognition elements that are as a result of their wider availability and specificity (Naresh and Lee, 2021).

One of the key processes involved in the application of biosensors is the gradual immobilization of the biological element at the surface of the transducer. There are different methods of achieving the immobilization step with the main commonly being solid phase adsorption, which is a physical process, intermolecular cross

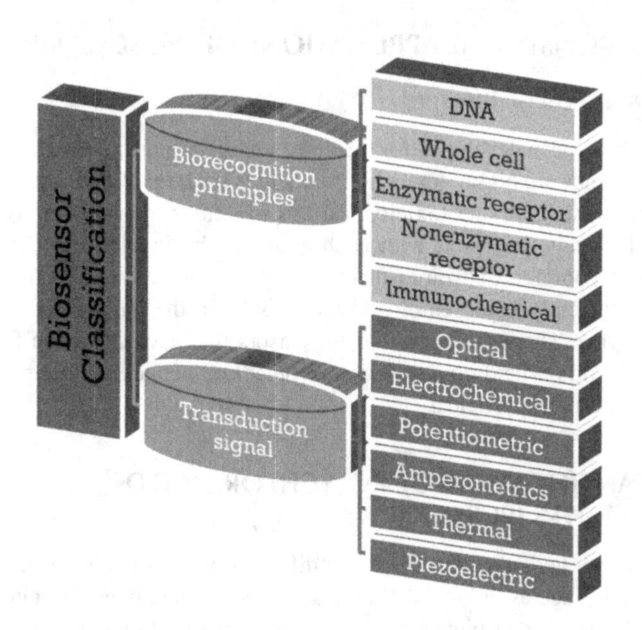

FIGURE 9.1 Classification of Biosensors.

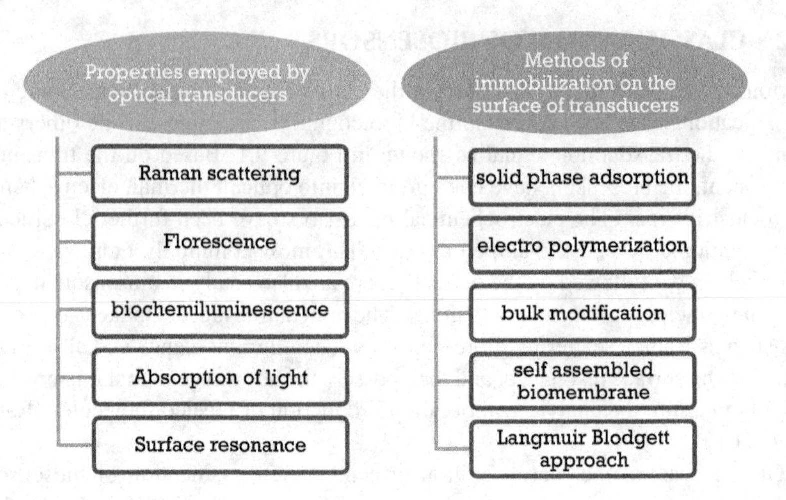

FIGURE 9.2 Properties Employed by Optical Transducers and Methods of Immobilization on the Surface of Transducers.

linking, surface adsorption through covalent bonding and entrapment with polymer, surfactant matrix or microcapsule. Besides the traditional approaches, other methods include: electro polymerization, bulk modification, self-assembled biomembrane and the Langmuir Blodgett approach to deposition (Umemura, 2018). The properties employed by optical transducers and methods of immobilization on the surface of transducers is shown in Figure 9.2.

9.3 ENVIRONMENTAL APPLICATIONS OF BIOSENSORS

9.3.1 Detection of Biocides

Biosensors have been developed for the detection of various biocidal agents in water. These compounds are highly toxic; hence, there is concern of their presence in water due to their persistence and toxicity. Various traditional approaches were formerly employed for their detection, with each approach having its inherent limitations. For the most common and rapid determination of these contaminants, enzymatic sensors are currently being employed and this is on the basis of the inhibition of selected enzyme. There are numerous biosensors that have also been fabricated for the detection and on-field quantification of other compounds such as carbamate and organophosphoruous pesticides.

9.4 QUANTIFICATION OF POLYCHLORINATED BIPHENYLS (PCBS)

These are compounds that are present in different parts of the environment because of their broad usage as industrial chemicals, most especially as capacitors and electrical transformers. The high toxic effects of these contaminants on humans and other animals make them of important concern. Although PCBs have been banned in various countries, their concentrations continue to rise (Victorious et al., 2019).

9.4.1 Detection of Dioxins

Aside PCBs, there are other polychlorinated compounds of serious concern. These include the dioxins, which are discharged as by-products in various chemical processes that involve chlorine. Dioxins are released during various processes involving the production of pulp, PVC plastics, some pesticides and incineration of waste. These compounds are known to be carcinogenic, hence, constitute serious threats to humans.

9.5 SURFACTANTS

There are various biosensors that have been developed for the detection of surfactants in various environmental matrices. The application of amperometric biosensors has also been well documented.

9.6 ANTIBIOTICS

Until recently, medical waste from drugs has been continually discharged into the environment indiscriminately. The increasing contents of antibiotics in the environment constitute a serious concern since most of these give rise to antibiotic resistance. The rising usage of antibiotics has given rise to a significant food safety concern.

9.7 BIOSENSORS WITH OTHER METHODS OF SEPARATION

Chemical separation processes are known to be very efficient, but some of the components to be separated tend to be too complex, making chromatography highly limited. Also, the use of biosensors has the challenge of poor selectivity in the detection of some of the contaminants. The two approaches could be integrated to give rise to better efficiency. In such combined approaches, the components are separated using separation retention time and functional recognition. Figure 9.3 shows environmental applications of biosensors

9.8 STUDIES ON BIOSENSORS FOR ENVIRONMENTAL APPLICATION

Odobašić et al. (2019) observed that biosensors are indispensable alternatives to the traditional analytical methods for the assessment and monitoring of water quality for human safety. They highlighted the role of biosensors in ensuring water quality with respect to pollution by heavy metals present in the environment.

Gutierrez et al. (2015) reviewed the various advantages and limitations of using eukaryotic microbes for the design of various types of whole-cell biosensing devices. In their work they took into consideration three major types of eukaryotic classification groups such as microalgae, yeasts and ciliated group of protozoa.

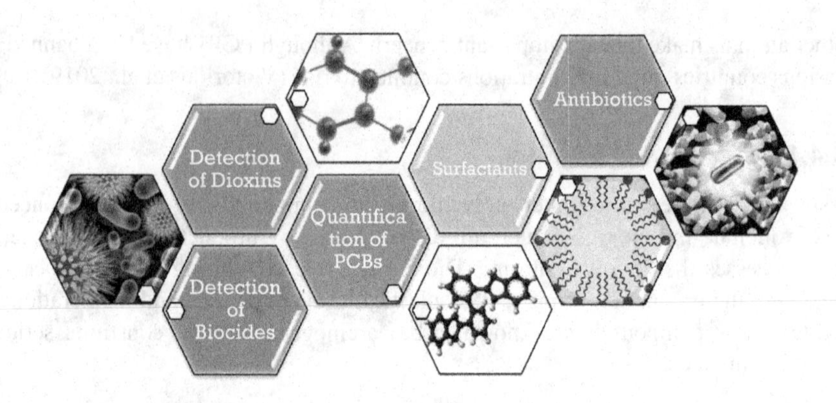

FIGURE 9.3 Environmental Applications of Biosensors.

Rodriguez-Mozaz et al. (2004) opined in their review work that biosensors are excellent tools for achieving various environmental monitoring programs for implementation of various regulations with respect to pollution. They discussed some of the aspects of biosensors, most especially with regard to miniaturization, nanotechnology and bioelectronics.

Hu et al. (2011) documented that environmental analysis and monitoring area remarkable premise of environment due to its provision for environmental protection and efficient monitoring. They also emphasized the need for online and on-site monitoring, which can be achieved through the use of biosensors. They therefore summarized the fabrication, features and types of biosensors that can be employed for environmental monitoring.

Tovar-Sanchez et al. (2019) in their study did a comprehensive analysis of bacteria, some plant species, small mammals and lichens as sensors for monitoring environmental metals. They also evaluated the use of various biomarkers on the device.

Ranveer et al. (2015) documented that biosensors have the potential of complementing the conventional approaches that are laboratory based in the area of environmental monitoring and control. They highlighted various interrelated issues that are connected with the use of biosensors from the regulatory perspectives.

9.9 CONCLUSION AND FUTURE TRENDS

There has been remarkable advancement in the area of biosensors for environmental monitoring. In spite of the various potentials of these miniature devices as well as the great number being fabricated, there is restricted application of commercial biosensors for environmental purpose. On a general note, some limitations are currently associated with the application of biosensors in the area of the environment: response time, sensitivity and lifetime. There is, therefore, need for further studies in this regard. There are also numerous areas which, when advanced, could further improve on the application of biosensors in the environment. Some include: integration with nanotechnology, utilization of immobilization approach during quantification, determinations using an array of multisensors, and miniaturization. The need for producing novel sensing components that are not difficult to fabricate will aid in widening the selectivity of the spectra. Furthermore, the development of advanced receptors will aid the quantification of complex environment samples as well as on-site analysis. There is, however, the need for creating more reliable and improved devices. There is need for the development of sensors that are able to determine different multiple analytes at the same time. This would be paramount in saving time, volume of sample, and the other reagents that are needed. There is a rising trend focusing on the aspect of miniaturization of analytical processes, which makes it easier for the handling of minute quantities of the sample. The future of biosensors therefore relies significantly on the achievement of developing complex micro- and nanotechnology, thin film science, biochemistry, and electronics (Adetunji et al., 2021a,b,c; Ukhurebor et al., 2021; Sangeetha et al., 2021; Adetunji and Ukhurebor, 2021; Oluwaseun et al., 2017; Adetunji et al., 2012; Arowora et al., 2012; Dauda et al., 2022a,b; Ukhurebor et al., 2021; Okeke et al., 2021; Adetunji and Anani, 2021; Nwankwo et al., 2021; Adejumo and Adetunji, 2018; Ukhurebor et al., 2021).

REFERENCES

Adejumo, I.O., & Adetunji, C.O. (2018). Production and evaluation of biodegraded feather meal using immobilised and crude enzyme from *Bacillus subtilis* on broiler chickens. *Brazilian Journal of Biological Sciences*, 5(10), 405–416.

Adetunji, C.O., & Anani, O.A. (2021). Bioaugmentation: A powerful biotechnological techniques for sustainable ecorestoration of soil and groundwater contaminants. In: Panpatte, D.G., & Jhala, Y.K. (eds.), Microbial Rejuvenation of Polluted Environment: Microorganisms for Sustainability, vol. 25. Singapore: Springer. https://doi.org/10.1007/978-981-15-7447-4_15

Adetunji, C.O., Anani, O.A., Olaniyan, O.T., Inobeme, A., Olisaka, F.N., Uwadiae, E.O., & Obayagbona, O.N. (2021a). Recent trends in organic farming. In: Soni, R., Suyal, D.C., Bhargava, P., & Goel, R. (eds.), *Microbiological Activity for Soil and Plant Health Management*. Singapore: Springer. https://doi.org/10.1007/978-981-16-2922-8_20

Adetunji, C.O., Egbuna, C., Oladosun, T.O., Akram, M., Michael, O., Olisaka, F.N., Ozolua, P., Adetunji, J.B., Enoyoze, G.E., & Olaniyan, O. (2021b). Efficacy of phytochemicals of medicinal plants for the treatment of human echinococcosis. Chapter 8. In: Egbuna, C., Akram, M., & Ifemeje, J.C. (eds.), *Neglected Tropical Diseases and Phytochemicals in Drug Discovery*. DOI: 10.1002/9781119617143

Adetunji, C.O., Fawole, O.B., Afolayan, S.S., Olaleye, O.O., & Adetunji, J.B. (2012). An Oral Presentation During 3rd NISFT Western Chapter Half Year Conference/General Meeting, Ilorin, pp. 14–16.

Adetunji, C.O., Michael, O. S., Nwankwo, W., Ukhurebor, K.E., Anani, O.A., Oloke, J.K., Varma, A., Kadiri, O., Jain, A., & Adetunji, J.B. (2021c). Quinoa, the next biotech plant: Food security and environmental and health hot spots. In: Varma, A. (ed.), *Biology and Biotechnology of Quinoa*. Singapore: Springer. https://doi.org/10.1007/978-981-16-3832-9_19

Adetunji, C.O., Mitembo, W.P., Egbuna, C., & Narasimha Rao, G.M. (2020). Silico modeling as a tool to predict and characterize plant toxicity. In: Mtewa, A.G., Egbuna, C., & Narasimha Rao, G.M. (eds.), *Poisonous Plants and Phytochemicals in Drug Discovery*. Wiley Online Library. https://doi.org/10.1002/9781119650034.ch14

Adetunji, C.O., Nwankwo, W., Olayinka, A.S., Olugbemi, O.T., Akram, M., Laila, U., Olugbenga, M.S., Oshinjo, A.M., Adetunji, J.B., Okotie, G.E., Esiobu, N. (Diuto). (2022a). Computational intelligence techniques for combating COVID-19. In: *Medical Biotechnology, Biopharmaceutics, Forensic Science and Bioinformatics*. CRC Press. Pages 12. eBook ISBN 9781003178903. DOI: 10.1201/9781003178903-16

Adetunji, C.O., Nwankwo, W., Olayinka, A.S., Olugbemi, O.T., Akram, M., Laila, U., Olugbenga, M.S., Oshinjo, A.M., Adetunji, J.B., Okotie, G.E., Esiobu, N. (Diuto). (2022b). Machine learning and behaviour modification for COVID-19. In: *Medical Biotechnology, Biopharmaceutics, Forensic Science and Bioinformatics*. CRC Press. Pages 17. eBook ISBN 9781003178903. DOI: 10.1201/9781003178903-17

Adetunji, C.O., & Oyeyemi, O.T. 2022. Antiprotozoal activity of some medicinal plants against *Entamoeba histolytica*, the causative agent of amoebiasis. In: *Medical Biotechnology, Biopharmaceutics, Forensic Science and Bioinformatics*. CRC Press. Pages 12. eBook ISBN 9781003178903. www.taylorfrancis.com/chapters/edit/10.1201/9781003178903-20/antiprotozoal-activity-medicinal-plants-entamoeba-histolytica-causative-agent-amoebiasis-charles-oluwaseun-adetunji-oyetunde-oyeyemi.

Adetunji, C.O., & Ukhurebor, K.E. (2021). Recent trends in utilization of biotechnological tools for environmental sustainability. In: Adetunji, C.O., Panpatte, D.G., & Jhala, Y.K. (eds.), *Microbial Rejuvenation of Polluted Environment: Microorganisms for Sustainability*, vol. 27. Singapore: Springer. https://doi.org/10.1007/978-981-15-7459-7_11

Adetunji, C.O.O., Olaniyan, T., Adeyomoye, O., Dare, A., Adeniyi, M.J., Alex, E., Rebezov, M., Garipova, L., & Ali Shariati, M. (2022c). eHealth, mHealth, and telemedicine for COVID-19 pandemic. In: Pani, S.K., Dash, S., dos Santos, W.P., Chan Bukhari, S.A., & Flammini, F. (eds.), *Assessing COVID-19 and Other Pandemics and Epidemics using Computational Modelling and Data Analysis*. Cham: Springer. https://doi.org/10.1007/978-3-030-79753-9_10

Adetunji, C.O.O., Olaniyan, T., Adeyomoye, O., Dare, A., Adeniyi, M.J., Alex, E., Rebezov, M., Isabekova, O., & Ali Shariati, M. (2022d). Smart sensing for COVID-19 pandemic. In: Pani, S.K., Dash S., dos Santos, W.P., Chan Bukhari, S.A., & Flammini F. (eds.), *Assessing COVID-19 and Other Pandemics and Epidemics using Computational Modelling and Data Analysis*. Cham: Springer. https://doi.org/10.1007/978-3-030-79753-9_9

Adetunji, C.O.O., Olaniyan, T., Adeyomoye, O., Dare, A., Adeniyi, M.J., Alex, E., Rebezov, M., Koriagina, N., & Ali Shariati, M. (2022e). Diverse techniques applied for effective diagnosis of COVID-19. In: Pani, S.K., Dash, S., dos Santos, W.P., Chan Bukhari, S.A., & Flammini, F. (eds.), *Assessing COVID-19 and Other Pandemics and Epidemics using Computational Modelling and Data Analysis*. Springer, Cham. https://doi.org/10.1007/978-3-030-79753-9_3

Adetunji, C.O.O., Olaniyan, T., Adeyomoye, O., Dare, A., Adeniyi, M.J., Alex, E., Rebezov, M., Petukhova, E., & Ali Shariati, M. (2022f). Machine learning approaches for COVID-19 pandemic. In: Pani, S.K., Dash, S., dos Santos, W.P., Chan Bukhari, S.A., & Flammini, F. (eds.), *Assessing COVID-19 and Other Pandemics and Epidemics using Computational Modelling and Data Analysis*. Cham: Springer. https://doi.org/10.1007/978-3-030-79753-9_8

Adetunji, C.O.O., Olaniyan, T., Adeyomoye, O., Dare, A., Adeniyi, M.J., Alex, E., Rebezov, M., Petukhova, E., & Ali Shariati, M. (2022g). Internet of health things (IoHT) for COVID-19. In: Pani, S.K., Dash, S., dos Santos, W.P., Chan Bukhari, S.A., & Flammini, F. (eds.), *Assessing COVID-19 and Other Pandemics and Epidemics using Computational Modelling and Data Analysis*. Cham: Springer. https://doi.org/10.1007/978-3-030-79753-9_5

Adetunji, C.O.O., Olugbemi, T., Akram, M., Laila, U., Samuel, M.O., Oshinjo, A.M., Adetunji, J.B., Okotie, G.E., Esiobu, N. (Diuto), Oyedara, O.O., & Adeyemi, F.M. (2022h). Application of computational and bioinformatics techniques in drug repurposing for effective development of potential drug candidate for the management of COVID-19. In: *Medical Biotechnology, Biopharmaceutics, Forensic Science and Bioinformatics*. CRC Press. Pages 14. eBook ISBN 9781003178903. DOI: 10.1201/9781003178903-15

Akram, M., Adetunji, C.O., Egbuna, C., Jabeen, S., Olaniyan, O., Ezeofor, N.J., Anani, O.A., Laila, U., Găman, M.-A., Patrick-Iwuanyanwu, K., Ifemeje, J.C., Chikwendu, C.J., Michael, O.C., & Rudrapal, M. (2021a). Dengue fever, Ch17. In: *Neglected Tropical Diseases and Phytochemicals in Drug Discovery*. Wiley. DOI: 10.1002/9781119617143

Akram, M., Mohiuddin, E., Adetunji, C.O., Oladosun, T.O., Ozolua, P., Olisaka, F.N., Egbuna, C., Michael, O., Adetunji, J.B., Hameed, L., Awuchi, C.G., Patrick-Iwuanyanwu, K., & Olaniyan, O. (2021b). Prospects of phytochemicals for the treatment of helminthiasis, Ch7. In: *Neglected Tropical Diseases and Phytochemicals in Drug Discovery*. Wiley. DOI: 10.1002/9781119617143

Arowora, K.A., Abiodun, A.A., Adetunji, C.O., Sanu, F.T., Afolayan, S.S., & Ogundele, B.A. (2012). Levels of aflatoxins in some agricultural commodities sold at Baboko Market in Ilorin, Nigeria. *Global Journal of Science Frontier Research*, 12(10), 31–33.

Dauda, W.P., Abraham, P., Glen, E., Adetunji, C.O., Ghazanfar, S., Ali, S., Al-Zahrani, M., Azameti, M.K., Alao, S.E.L., Zarafi, A.B., Abraham, M.P., & Musa, H. (2022a). Robust profiling of cytochrome P450s (P450ome) in notable Aspergillus spp. *Life*, 12(3), 451. https://doi.org/10.3390/life12030451

Dauda, W.P., Morumda, D., Abraham, P., Adetunji, C.O., Ghazanfar, S., Glen, E., Abraham, S.E., Peter, G.W., Ogra, I.O., Ifeanyi, U.J., Musa, H., Azameti, M.K., Paray, B.A., & Gulnaz, A. (2022b). Genome-wide analysis of cytochrome P450s of *Alternaria* species: Evolutionary origin, family expansion and putative functions. *Journal of Fungi*, 8(4), 324. https://doi.org/10.3390/jof8040324

Gavrilaş, S., Ursachi, C.Ş., Perţa-Crişan, S., & Munteanu, F.-D. (2022). Recent trends in biosensors for environmental quality monitoring. *Sensors (Basel)*. February, 22(4), 1513. DOI: 10.3390/s22041513

Gutiérrez, J.C., Amaro, F., & Martín-González, A. (2015). Heavy metal whole-cell biosensors using eukaryotic microorganisms: An updated critical review. *Frontier Microbiology*, 20 February, 6, 48. https://doi.org/10.3389/fmicb.2015.00048

Hu, N., Zhao, M.-H., Chuang, J.-L., & Li, L. (2011). The application study of biosensors in environmental monitoring. In: *Proceedings of 2011 Cross Strait Quad-Regional Radio Science and Wireless Technology Conference*, 26–30 July 2011, IEEE, Harbin, China.

Jain, P., Coogn, S., & Flannigan, M. (2020). A review of machine learning applications in wildfire science and management. *Environmental Reviews*. https://doi.org/10.1139/er-2020-001

Kendler, S., & Fishbain, B. (2022). Optimal wireless distributed sensor network design and ad-hoc deployment in a chemical. *Sensors*, 22(7), 2563. https://doi.org/10.3390/s22072563

Li, R. (2020). A review of remote sensing for environmental monitoring in China. *Remote Sensing*, 12, 1130. https://doi:10.3390/rs12071130

Lin Q., Zhag F., Jiang W., & Wu, H. (2018). Environmental monitoring of ancient buildings based on a wireless sensor network. *Sensors*, 18(12), 4234. https://doi.org/10.3390/s18124234

Mitchell, A.L., Rosenqvist, A., & Mora, B. (2017). Current remote sensing approaches to monitoring forest degradation in support of countries measurement, reporting and verification (MRV) systems for REDD+. *Carbon Balance Manage*, 12(1), 9. https://doi.org/10.1186/s13021-017-0078-9

Odobašić, A.S., Šestan, I., & Begić, S. (2019). Biosensors for determination of heavy metals in waters. In: *Biosensors for Environmental Monitoring*. IntechOpen. DOI: 10.5772/intechopen.84139

Okeke, N.E., Adetunji, C.O., Nwankwo, W., Ukhurebor, K.E., Makinde, A.S., & Panpatte, D.G. (2021). A critical review of microbial transport in effluent waste and sewage sludge treatment. In: Adetunji, C.O., Panpatte, D.G., & Jhala, Y.K. (eds.), Microbial Rejuvenation of Polluted Environment: Microorganisms for Sustainability, vol. 27. Singapore: Springer. https://doi.org/10.1007/978-981-15-7459-7_10

Oluwaseun, A.C., Phazang, P., & Sarin, N.B. (2017). Significance of rhamnolipids as a biological control agent in the management of crops/plant pathogens. *Current Trends in Biomedical Engineering & Biosciences,* 10(3), 54–55.

Olugbemi, O. T., Adetunji, C.O., Adeniyi, M.J., & Ingo Hefft, D. (2022). Machine learning techniques for high-performance computing for IoT applications in healthcare. In: *Deep Learning, Machine Learning and IoT in Biomedical and Health Informatics*. CRC Press. Pages 13. eBook ISBN 9780367548445. DOI: 10.1201/9780367548445-20 2022

Olugbemi, O.T., Adetunji, C.O., Adeniyi, M.J., & Ingo Hefft, D. (2022). Computational Intelligence in IoT Healthcare. In: *Deep Learning, Machine Learning and IoT in Biomedical and Health Informatics*. CRC Press. Pages 13. eBook ISBN 9780367548445. DOI: 10.1201/9780367548445-19

Oyedara, O.O., Adeyemi, F.M., Adetunji, C.O., & Elufisan, T.O. (2022). Repositioning antiviral drugs as a rapid and cost-effective approach to discover treatment against SARS-CoV-2 infection. In: *Medical Biotechnology, Biopharmaceutics, Forensic Science and Bioinformatics.* CRC Press. Page 12. eBook ISBN 9781003178903. DOI: 10.1201/9781003178903-10

Parthipan, V., & Dhanasekaran, D. (2019). Preventing and monitoring of framework for forest fire detection and data analysis using internet of things (IoT). *International Journal of Engineering and Advanced Technology (IJEAT),* 8(3S). ISSN: 2249-8958.

Phumlani, T., Shumbula, P.M., & Njengele-Tetyana, Z. (2010). *Biosensors: Design, Development and Applications.*

Ranveer, A., Nasipude, S., & Bagwan, S. (2015). Biosensor for environmental monitoring. *International Journal of Innovations in Engineering Research and Technology* (IJIERT), 2(4), 1–8. ISSN: 2394-3696.

Rodriguez-Mozaz, S., Marco, M.-P., Lopez de Alda, M.J., & Barceló, D. (2004). Biosensors for environmental applications: Future development trends. *Pure and Applied Chemistry,* 76(4), 723–752. IUPAC723.

Sangeetha, J., Hospet, R., Thangadurai, D., Adetunji, C.O., Islam, S., Pujari, N., & Al-Tawaha, A.R.M.S. (2021). Nanopesticides, nanoherbicides, and nanofertilizers: The greener aspects of agrochemical synthesis using nanotools and nanoprocesses toward sustainable agriculture. In: Kharissova, O.V., Torres-Martínez, L.M., & Kharisov, B.I. (eds.), *Handbook of Nanomaterials and Nanocomposites for Energy and Environmental Applications*. Cham: Springer. https://doi.org/10.1007/978-3-030-36268-3_44

Tovar-Sánchez, E., Suarez-Rodríguez, R., Ramírez-Trujillo, A., Valencia-Cuevas, L., Hernández-Plata, I., & Mussali-Galante, P. (2019). The use of biosensors for biomonitoring environmental metal pollution. DOI: 10.5772/intechopen.84309

Ukhurebor, K.E., Adetunji, C.O., Bobadoye, A.O., Aigbe, U.O., Onyancha, R.B., Siloko, I.U., Emegha, J.O., Okocha, G.O., & Abiodun, I.C. (2021). *Bionanomaterials for biosensor technology.* Bionanomaterials. Fundamentals and Biomedical Applications, pp. 5–22.

Ukhurebor, K.E., Mishra, P., Mishra, R.R., & Adetunji, C.O. (2020). Nexus between climate change and food innovation technology: Recent advances. In: Mishra, P., Mishra, R.R., & Adetunji, C.O. (eds.), *Innovations in Food Technology*. Singapore: Springer. https://doi.org/10.1007/978-981-15-6121-4_20

Ukhurebor, K.E., Nwankwo, W., Adetunji, C.O., & Makinde, A.S. (2021). Artificial intelligence and internet of things in instrumentation and control in waste biodegradation plants: Recent developments. In: Adetunji, C.O., Panpatte, D.G., & Jhala, Y.K. (eds.), *Microbial Rejuvenation of Polluted Environment: Microorganisms for Sustainability*, vol 27. Singapore: Springer. https://doi.org/10.1007/978-981-15-7459-7_12

Umemura, Y. (2018). Preparation and application of clay mineral films. *Developments in Clay Science*, 9, 377–396. https://doi.org/10.1016/B978-0-08-102432-4.00012-3

Varnakavi, N., & Lee, N. (2021). A review on biosensors and recent development of nanostructured materials-enabled biosensors. *Sensors (Basel)*, 21(4), 1109. Published online 2021 Feb 5. DOI: 10.3390/s21041109

Victorious, A., Saha, S., Pandey, R., Didar, T.F., & Soleymani, L. (2019). Affinity-based detection of biomolecules using photo-electrochemical readout. *Frontiers in Chemistry*, 11 September. 10.3389/fchem.2019.00617

Yang, Y., & Deng, Z. (2019). Stretchable sensors for environmental monitoring. *Applied Physics Reviews*, 6, 011309. https://doi.org/10.1063/1.5085013

10 Artificial Intelligence and Its Role in the Food Industry

Akinola Samson Olayinka, Onyijen Ojei Harrison, Tosin Comfort Olayinka, Charles Oluwaseun Adetunji, and Tsegaye Bojago Dado

CONTENTS

10.1 INTRODUCTION TO ARTIFICIAL INTELLIGENCE (AI)

AI technology has become the technology of the future as it was in the past decade. (Boucher, 2020; Kakani et al., 2020; Misra et al., 2020(Soltani-Fesaghandis & Pooya, 2018, (Habeeb, 2017).

Rather than adding to the rising number of AI definitions, today's definitions usually include additional requirements such as autonomy and the capacity to limit intelligence to specific domains (Samoili et al., 2020). The European Union department for communication defines artificial intelligence as "systems that display intelligent behavior by assessing their environment and acting independently to attain particular goals" (ECC, 2018). AI is the science and engineering of building intelligent machines, particularly clever computer programs, according to John McCarthy, known as the "father" of AI (Donepudi, 2014). Machine learning and deep learning are two of the most extensively used branches in the field of AI. These branches are used by individuals, corporations, and government agencies to carry out predictions based on available data. Machine learning models are increasingly being created to deal with the complexity and variety of data in the food industry (Vadlamudi, 2019).

DOI: 10.1201/9781003207955-10

The deployment of artificial intelligence in various spheres of life has produced desired positive and significant results in recent times (Misra et al., 2020). Leveraging AI technology has fulfilled social requirements while ensuring timely delivery of high-quality goods in manufacturing, such as food. Using new technologies, the food industry might produce a large number of food items in less time, boosting a company's food manufacturing output and revenue (Misra et al., 2020). Artificial intelligence is used in essentially every sector of technology and enables people globally to solve problems quickly, computerize the food sector, and modify food items (Soltani-Fesaghandis & Pooya, 2018). AI enhances the way the food industry handles seed selection, crop monitoring, watering, and temperature monitoring to meet required agricultural standards, resulting in superior food industry products (Donepudi, 2014; Vadlamudi, 2019).

In the food industry, where developing standard, reliable procedures to control product quality is a major goal, the search for new ways to reach and serve customers while keeping costs low necessitated the use of AI to improve the customer experience, supply chain management, operational efficiency, material movement, and vehicle activity reduction. (Kumar et al., 2021)

10.2 TYPES OF ARTIFICIAL INTELLIGENCE

Artificial Intelligence can be divided into several categories. Type I has no memory and can only respond to the current stimulus. Type II only has a small amount of memory to work with while making decisions. Type III will help with memory and decision-making. Type IV AI refers to a type of AI that is aware of both its own and others' feelings. Each of the four categories has two components: learning and decision-making, which are characterized as follows and shown in Figure 10.1 (Johnson, 2020).

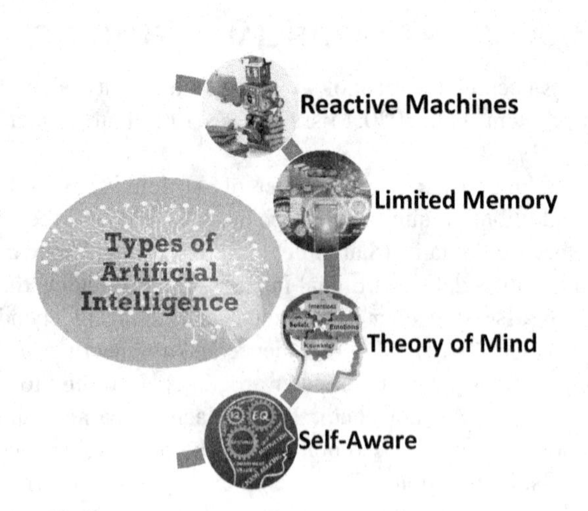

FIGURE 10.1 Type of Artificial Intelligence.

Reactive Machines: These are used only to execute simple tasks such as face detection and are a basic level of AI. These types provide an output in response to some input. Any AI system's first stage, is to detect a human face. A simple reactive system inputs the image of a human face and creates a box around it. No information is saved or learned by the model. Machine learning models that are static are reactive machines. Their architecture is the most basic, and they are available on repositories such as GitHub all over the Internet. It is simple to download, trade, pass around, and import these models into a developer's toolbox.

Limited Memory: It is the ability of this type of AI to recall past facts and forecasts in order to make more accurate predictions. Machine learning architecture becomes a little more challenging when memory is limited. Despite the fact that every machine-learning model requires a small amount of memory to build; it can be employed as a reactive machine. Reinforcement learning, long short-term memory, and evolutionary generative adversarial networks are examples of machine-learning models that achieve this limited memory type.

Theory of Mind: This is a level of AI that is not yet fully achieved. They are still at the stage of development such as self-driving cars. They are built to have interactions with the emotions and thought of humans involved.

Self-Aware: AI is currently designed to become self-aware. Only in fiction can this type of AI exist, and it fills viewers with both hope and terror, as stories frequently do. As a self-aware intelligence beyond human cognition has its own intelligence, humanity will very probably have to negotiate conditions with the creature that is generated.

10.3 COMPONENTS OF ARTIFICIAL INTELLIGENCE

Figure 10.2 shows the various components of artificial intelligence and the different components are discussed as follows:

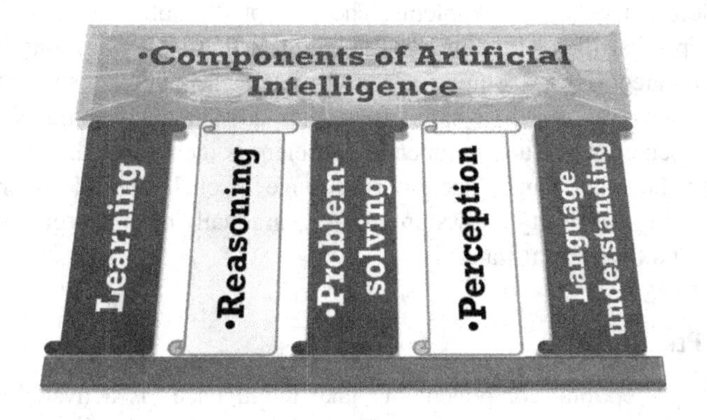

FIGURE 10.2 Components of Artificial Intelligence.

10.3.1 Learning

Computer programs, like humans, learn in a variety of ways. Trial-and-error, computer algorithm and generalization methods are ways computer programs can learn. The trial-and-error method is one of the vital methods of AI learning, which continues to solve problems until it reaches the desired results. As a result, the program maintains track of all the steps that resulted in positive outcomes and saves them in its database for use when the computer faces the same situation again. AI, like humans, must first learn a task before performing it. When confronted with similar or identical facts, the human brain organizes information so that it can make quick judgements in the future. A computer algorithm, on the other hand, constructs a model of thinking with parameters that are dependent on external data input. It is a method for resolving future problems quickly and accurately. Individual objects, such as distinct issue solutions, vocabulary, foreign languages, and so on, are memorized as part of AI's learning component. The generalization method is then used to put this learning into practice (Bansal, 2021). There are numerous machine-learning categories, including supervised, unsupervised, and reinforcement learning.

10.3.2 Reasoning

One of the most important properties of artificial intelligence is its capacity to differentiate. Allowing the platform to reason entails allowing it to draw conclusions that are appropriate for the situation. Inductive and deductive inferences are two types of inferences. The difference is that in an inferential scenario, a problem's solution ensures a result. In the inductive inference, however, the accident is always caused by instrument failure. In computer programming, deductive interferences have had a lot of success. On the other hand, drawing meaningful inferences from the current circumstance is always a component of the reasoning process.

10.3.3 Problem Solving

AI's ability for problem-solving is based on availability of data. The AI platform is used to address a variety of problems. The multiple "problem-solving" approaches are made up of artificial intelligence components that split questions into *special* and *general* categories. A solution to a given problem is tailor-made in the case of a special-purpose approach, which often takes advantage of some of the unique qualities presented when a recommended problem is incorporated. A multi-purpose strategy, on the other hand, faces a wide range of challenges. Furthermore, AI's problem-solving capability allows programs to gradually close the gap between any target state and the current state.

10.3.4 Perception

When the "perception" component of "fake intelligence" is activated, it analyzes any given surrounding with a variety of artificial or real sense organs. Furthermore the processes are kept internal, allowing the perceiver to investigate numerous circumstances in suggested items and to comprehend their connections and features. Because similar products can have a range of looks depending on the angle from which they are seen, this research is frequently difficult. Perception is one of the components of artificial intelligence that can be used to propel self-driving cars at low speeds.

10.3.5 Language Understanding

Language is made up of a variety of system indications that follow a set of rules to justify their behavior. Language understanding, which utilizes multiple sorts of language across diverse forms of natural meaning, is one of the most extensively used artificial intelligence components. AI is trained to understand English, which is the most widely spoken human language. As a result, the platform enables computers to swiftly comprehend the many programs that they run (Bansal, 2021).

10.4 BIG DATA IN AI AND ITS ROLE IN THE FOOD INDUSTRY

Big data is defined by the World Health Organization as rapidly collected and complex data in vast numbers (Ward & Barker, 2013). Big data has three main features, according to Katal et al. (2013): a great amount of data, a quick rate of change, and a wide range of information. The European Commission (EC) defines big data as huge volumes of data gathered from a variety of sources (EC, 2022). New tools and technology, such as robust algorithms, are necessary to deal with such rapidly changing data. Volume refers to the amount of data in all of the above definitions, which in the case of big data could be measured in terabytes. Velocity refers to the rate at which data is generated as well as the rate at which it should be analyzed and acted upon. The numerous kinds of data available, such as structured, semi-structured, and unstructured data, are referred to as variety (Aurthur, 2013). A dataset is said to have great variety if it has structural heterogeneity. Structured data is tabular data seen in spreadsheets or relational databases, accounting for only 10–15 percent of total data. Text, images, music, and video are examples of unstructured data. The Extensible Markup Language (XML), a textual language for exchanging data over the Internet, is used to express semi-structured data. In XML documents, user-defined data tags make the documents machine-readable. Semi-structured data accounts for around 5–10 percent of all data.

Due to the intrinsic features of big data, it is hard to process and utilize large and complicated datasets using standard data-management procedures. As a result, new and creative computing approaches are necessary for the gathering, storage, dissemination, analysis, and management of large data (Geczy, 2015; Lazer et al., 2014). Big data analytics refers to the methodologies for gathering, analyzing, and evaluating large datasets. One of the most commonly mentioned benefits of the big data revolution is the capacity to extract usable knowledge and feasible patterns from data (Jagadish et al., 2014; Mayer-Schonberger & Cukier, 2013). Some of the technologies and tools utilized in big data analytics include statistical analysis, data mining, data visualization, text analytics, social network analysis, signal processing, and machine learning (Chen & Zhang, 2014). Big data technology's purpose is to harness the power of massive amounts of data in real-time or otherwise (Daniel,

2019). The widespread usage of digital devices such as smartphones and sensors has resulted in the generation of an unprecedented volume of data (Gandomi & Haider, 2015).

Machine learning is used to create AI, which feeds data into an algorithm that learns answers. Large data is an important source of data for machine learning, but the degree of their potential contribution is unknown because big data often consists of unstructured free text. Machine learning is a branch of artificial intelligence that focuses on creating computer systems that can learn from and adapt to data without having to be explicitly programmed (Jordan & Mitchell, 2015). Machine learning algorithms can provide new insights, predictions, and solutions that are personalized to the needs and circumstances of each individual. When a large amount of high-quality input training data is available, machine learning techniques can produce reliable results and enhance informed decision-making (Gobert & Pedro, 2017; Gobert et al., 2012; Manyika et al., 2011) These data-intensive machine learning approaches are at the intersection of big data and artificial intelligence, and they have the potential to increase food production services and productivity.

In the food production process, sensors and other equipment are utilized to monitor food safety and give data for traceability. Records from the food processing, food transportation, and food retail industries are then kept for data processing. A huge amount of data is collected during these stages, and AI and big data technologies are useful for data processing. AI's goal is for computers to be able to "understand" what data means. Successful applications in computer vision, natural language processing, and other domains have appeared in recent years (Zhang et al., 2019). One of the most obvious advantages of AI is that it can acquire knowledge from a large amount of data automatically. As a result, AI requires a lot of data. The majority of big data approaches are used for collecting, storing, querying, and basic processing of large amounts of data. The information gathered has several repetitions. As a result, important information can be recovered from duplicated data using AI and big data.

In the food-manufacturing industry, AI assists in the monitoring of each stage of the process, forecasting price and inventory management, and following the path of things from where they are grown to where they are received by consumers, ensuring transparency. To avoid obtaining an overflow of commodities that will be squandered, food manufacturers can utilize technology like Symphony Retail AI to estimate demand for transportation, pricing, and inventory. Previously, a producer would have had to hire a huge number of people to perform the repetitive and monotonous activities of food selection. Instead of manually sorting large quantities of food by size and form in order to can or bag it, AI-based solutions can now instantly distinguish which plants should be used for potato chips and which for French fries (Kovalenko & Chuprina, 2022). The same technology will pick out vegetables with an unappealing color, minimizing the likelihood of customers abandoning them. Food sorters and peelers from TORMA have increased processing capacity and availability, improving food quality and safety. This is performed using core sensor technologies and a camera that recognizes material based on color, biological features, and form (length,

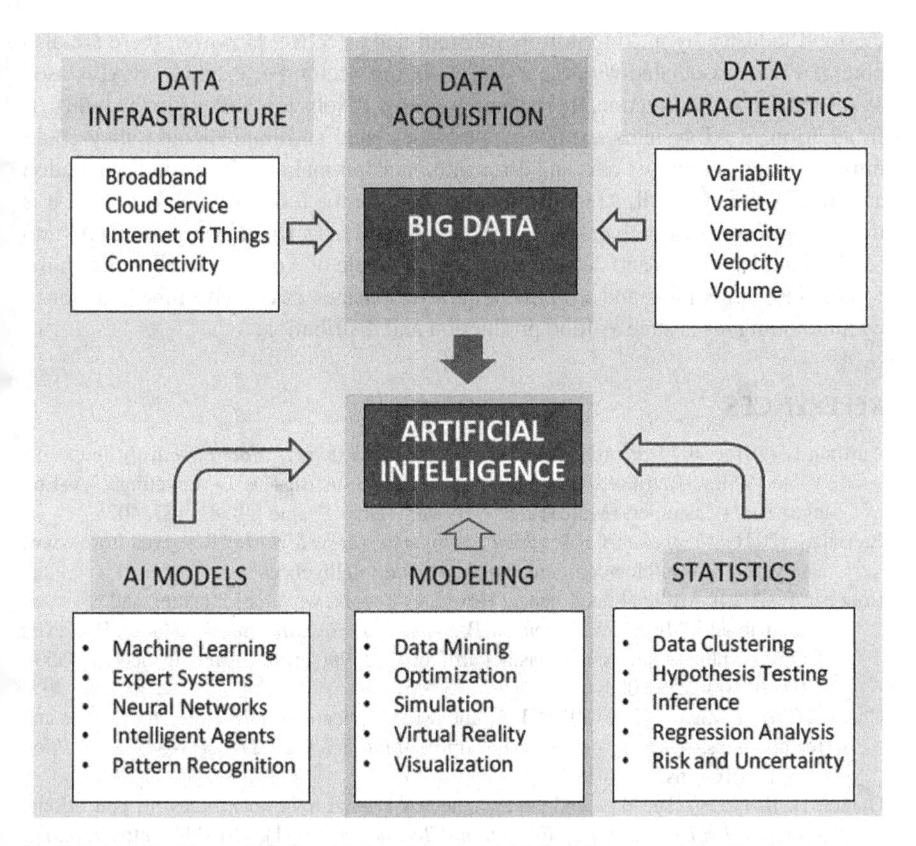

FIGURE 10.3 Shows Relationship between Artificial Intelligence and Big Data.

width, and diameter); the adaptive spectrum of the camera is particularly suited for optical food sorting (Kovalenko & Chuprina, 2022). Figure 10.3 shows the relationship between artificial intelligence and big data.

10.5 CONCLUSION

The role of Artificial Intelligence (AI) in the food industry is still in its infancy, with much potential yet to be explored. However, AI is already starting to have a significant impact on the sector, from agricultural production to food waste reduction. In the future, AI is likely to play an even bigger role in the food industry, with innovations such as intelligent packaging and personalized nutrition becoming a reality. As such, it is important to monitor the developments of AI in this field and to ensure that its application benefits consumers and producers alike. AI technologies can be used in various stages of the food supply chain, from farm to fork. For example, AI can be used in precision agriculture to optimize crop production, or in food processing to identify defects or contaminants. In addition, AI can be used in retail settings to track inventory and suggest recipes to consumers. AI has the potential to transform the food industry by making it more efficient and effective. However, there are also potential risks associated with the use of AI in this sector. For example, if AI is used to automate food production, there could be a loss of jobs for workers in the industry. In addition, if AI systems are not properly designed and implemented, they could introduce new sources of bias and error into decision-making about food production and distribution. Overall, AI holds great promise for the food industry. However, it is important to consider both the potential benefits and risks when incorporating AI into food-related processes and decision-making. The role of Artificial Intelligence in the food industry is evident and with the help of AI, businesses can save time and money by automating tasks such as food production and distribution.

REFERENCES

Aurthur, L. (2013). *Big Data Marketing – Engage Your Customers More Effectively and Drive Value*. Wiley. Retrieved from www.wiley.com/en-us/Big+Data+Marketing%3A+Engage+Your+Customers+More+Effectively+and+Drive+Value-p-9781118734025

Bansal, S. (2021). *Components of Artificial Intelligence – How It Works?* Retrieved from www.analytixlabs.co.in/blog/components-of-artificial-intelligence/

Boucher, P. (2020). Artificial intelligence: How does it work, why does it matter, and what can we do about it? In *EPRS: European Parliamentary Research Service* (June). Retrieved from https://policycommons.net/artifacts/1336906/artificial-intelligence/1944453/. CID: 20.500.12592/00dvnz.

Chen, P. C. L., & Zhang, C. Y. (2014). Data-intensive applications, challenges, techniques and technologies: A survey on Big Data. *Information Sciences*, *275*, 314–347. https://doi.org/10.1016/j.ins.2014.01.015

Daniel, B. K. (2019). Big data and data science: A critical review of issues for educational research. *British Journal of Educational Technology*, *50*(1), 101–113. https://doi.org/10.1111/bjet.12595

Donepudi, P. K. (2014). Technology growth in shipping industry: An overview. *American Journal of Trade and Policy*, *1*(3), 137–142. https://doi.org/10.18034/ajtp.v1i3.503

EC. (2022). Big data | Shaping Europe's digital future. Retrieved from https://digital-strategy.ec.europa.eu/en/policies/big-data

ECC. (2018). A European approach to artificial intelligence, MEMO. Google Search.

Gandomi, A., & Haider, M. (2015). Beyond the hype: Big data concepts, methods, and analytics. *International Journal of Information Management*, *35*(2), 137–144. https://doi.org/10.1016/j.ijinfomgt.2014.10.007

Geczy, P. (2015). Big data management: Relational framework. *Review of Business and Finance Studies*, *6*(3), 21–30.

Gobert, J. D., & Pedro, M. A. S. A. O. (2017). Digital assessment environments for scientific inquiry practices. *The Wiley Handbook of Cognition and Assessment: Frameworks, Methodologies, and Applications*. Wiley, pp. 508–534.

Gobert, J.D., Pedro, M. A. S. A. O., Baker, R. S. J. D., Toto, E., & Montalvo, O. (2012). Leveraging educational data mining for real-time performance assessment of scientific inquiry skills within Microworlds. *JEDM Special Issue*, *4*(3), 1–40.

Habeeb, A. (2017). Artificial intelligence. Ahmed Habeeb University of Mansoura. *Research Gate*, *7*(2). https://doi.org/10.13140/RG.2.2.25350.88645/1

Jagadish, H. V, Gehrke, J., Labrinidis, A., Papakonstantinou, Y., Patel, J. M., & Ramakrishnan, R. (2014). Big data and its technical challenges. *Communications of the ACM*, *57*(7), 2–9.

Johnson, J. (2020). *4 Types of Artificial Intelligence – BMC Software | Blogs*.

Jordan, M. I., & Mitchell, T. M. (2015). Machine learning: Trends, perspectives, and prospects. *Science*, *349*(6245), 255–260. https://doi.org/10.1126/science.aaa8415

Kakani, V., Nguyen, V. H., Kumar, B. P., Kim, H., & Pasupuleti, V. R. (2020). A critical review on computer vision and artificial intelligence in food industry. *Journal of Agriculture and Food Research*, 2(November 2020), 100033. https://doi.org/10.1016/j.jafr.2020.100033

Katal, A., Wazid, M., & Goudar, R. H. (2013). Big data: Issues, challenges, tools and good practices. *2013 6th International Conference on Contemporary Computing, IC3 2013*. https://doi.org/10.1109/IC3.2013.6612229

Kovalenko, O., & Chuprina, R. (2022). *Machine Learning and AI in Food Industry Solutions and Potential*. Kovalenko, SPD-Group. Retrieved from https://spd.group/machine-learning/machine-learning-and-ai-in-food-industry.

Kumar, I., Rawat, J., Mohd, N., & Husain, S. (2021). Opportunities of artificial intelligence and machine learning in the food industry. *Journal of Food Quality*. https://doi.org/10.1155/2021/4535567

Lazer, D., Ryan, K., Gary, K., & Vespignani, A. (2014). The parable of Google flu: Traps in big data analysis. *Science*, *343*(March), 1203–1205.

Manyika, J., Chui, M., Brown, B., Bughin, J., Dobbs, R., Roxburgh, C., & Hung Byers, A. (2011). *Big Data: The Next Frontier for Innovation, Competition and Productivity*. McKinsey Global Institute (June). https://doi.org/APO-147446

Mayer-Schonberger, V., & Cukier, K. (2013). Big data: A revolution that will transform how we live, work, and think. *Angewandte Chemie International Edition*, *6*(11), 951–952.

Misra, N. N., Dixit, Y., Al-Mallahi, A., Bhullar, M. S., Upadhyay, R., & Martynenko, A. (2020). IoT, big data and artificial intelligence in agriculture and food industry. *IEEE Internet of Things Journal*, *9*(9), 6305–6324. https://doi.org/10.1109/jiot.2020.2998584

Samoili, S., López Cobo, M., Gómez, E., De Prato, G., Martínez-Plumed, F., & Delipetrev, B. (2020). *AI Watch – Defining Artificial Intelligence: Towards an Operational Definition and Taxonomy of Artificial Intelligence*. Joint Research Centre (European Commission). https://doi.org/10.2760/382730

Soltani-Fesaghandis, G., & Pooya, A. (2018). Design of an artificial intelligence system for predicting success of new product development and selecting proper market-product strategy in the food industry. *International Food and Agribusiness Management Review*, *21*(7), 847–864. https://doi.org/10.22434/IFAMR2017.0033

Vadlamudi, S. (2019). How artificial intelligence improves agricultural productivity and sustainability: A global thematic analysis. *Asia Pacific Journal of Energy and Environment*, *6*(2), 91–100.

Ward, J. S., & Barker, A. (2013). *Undefined by Data: A Survey of Big Data Definitions*. 1–2. Retrieved from https://arxiv.org/abs/1309.5821

Zhang, H., Hong, X., Zhou, S., & Wang, Q. (2019). Infrared image segmentation for photovoltaic panels based on Res-UNet. In *Lecture Notes in Computer Science (Including Subseries Lecture Notes in Artificial Intelligence and Lecture Notes in Bioinformatics)*, 11857 LNCS (October), 1–13. Cham: Springer. https://doi.org/10.1007/978-3-030-31654-9_52

11 Recent Advances in the Characterization and Application of Graphene in the Food Industry

Charles Oluwaseun Adetunji, Olugbemi T. Olaniyan, Inobeme Abel, Olorunsola Adeyomoye, John Tsado Mathew, and Olalekan Akinbo

CONTENTS

11.1 INTRODUCTION

The atomic structure of graphene is a carbon nanomaterial planar sheet of two dimensional layers, a single atom thick membrane with significant importance in diverse fields of science. Much research supporting the utilization and applications of graphene has spanned between quantum mechanics to polymer science. The dynamic structure and properties – mechanical, electric, thermal, chemical and physical – have led to an increase in global market value driven by so much progress being made in applications and production. The word *graphene* was initially introduced by Boehm et al. in 1986 by denoting polyacyclic aromatic hydrocarbon. Through research, it was discovered that the carbon atom in graphene molecules have covalent bonds with three adjacent carbon atoms. Sahu et al. (2021) reported several processes and

DOI: 10.1201/9781003207955-11

methodologies involved in the preparation of graphene, such as fermentation, acid hydrolysis, electrochemical reduction, top down method, pyrolysis, oxidation, bottom up, and chemical vapor deposition. Elham et al. (2016) revealed that graphene is an interesting material with specific surface area, high intrinsic mobility, high young modulus, thermal conductivity, good electrical conductivity, optical transmittance, and transparent electrode conduction. The molecular structure of graphene is made up of oxide sheets with a basal plane coated with epoxy and hydroxyl functional groups and organic carboxylic acids. The authors revealed that the applications of graphene have been studied in polymer composites, sensors, gas detection, electronics, energy-related materials, paper-like materials, biomedical and field-effect transistors owing to the unique physiochemical properties. Elham et al. (2016) reported that several methods are available for the synthesis of graphene, such as chemical vapor deposition and high temperature–high pressure techniques, silicon carbide sublimation, cleavage, chemical oxidation of graphite and unfolding carbon nanotubes, which constitute a scalable, simple, and cheap approach. Characterization of graphene can be carried out through XRD theory, scanning electron microscopy, transmission electron microscopy and Raman spectra. Wang et al. (2010) demonstrated that graphene and its derivatives in recent times have gained unique attention across various fields in science, which is regarded as a mile stone in material science. Thus, this chapter focuses on providing an overview on the application, synthesis, and characterization of graphene.

11.2 SYNTHESIS AND PRODUCTION OF GRAPHENE

The term *graphene* is formed from two words; *graphite* (a carbon allotrope with hexagonally arranged atoms), and *ene* (unsaturated hydrocarbon). Graphene is a carbon allotrope made up of a two-dimensional hexagonal array of a single layer of atoms. It is characterized by high surface area, superb thermal, magnetic, mechanical and electronic features. As far as delivery of drugs and power storage are concerned, the production of graphene that exhibits manageable electronic and mechanical attributes is vital. Studies abound on the synthesis and production of graphene. Jiang et al. (2014) reported that preparation of graphene oxide (GO) nanosheets was executed through modified Hummers and Offeman. Then characterization of the capped oleic acid graphene oxide nanosheets was done using X-ray diffraction, transmission electron microscopy and Fourier transform infrared spectrophotometry. Mao et al. (2013) and Bhuyan et al. (2016) identified various methods of producing graphene. The methods include exfoliation of graphene by mechanical, chemical and thermal reduction, chemical vapor deposition and many more. They also noted the advantages and disadvantages.

11.3 EXFOLIATION

Exfoliation occurs whenever the rate of decomposition of epoxy and hydroxyl domains of graphite oxide overtakes the rate of diffusion of the evolved gases. This then produces pressures that are higher than the van der Waals bonds holding the graphene sheets (McAllister et al., 2007). Graphene materials can be obtained through exfoliation by electrochemical, mechanical, thermal, or electrical means.

11.4 ELECTROCHEMICAL EXFOLIATION

Electrochemical exfoliation can be described as a technique for the synthesis of graphene from graphites. It involves application of voltage to intercalate ionic species into graphites with the formation of species that exfoliate the individual graphene sheets. Operational preparation requirements include graphite rods, working electrodes and counter electrodes. One of the challenges that stand against the method includes the occurrence of graphite electrode disintegration, especially in the course of the process (Achee et al., 2018).

Fei et al. (2019) reviewed studies conducted between 2015 and 2019 on the synthesis of graphene materials with elevated carbon to oxygen (C/O) ratios from the perspectives of equipment engineering, electrolytes, graphite electrodes and extra reduction methods. The authors looked at the use of electrochemical techniques in doping of graphene materials, incorporation of different nanoparticles, and covalent functional group introduction, and then compared the synthesized graphene materials' properties. Among the properties compared are layer numbers, quality, lateral size and carbon to oxygen ratio.

Hofmann et al. (2015) examined electrochemical exfoliation process and the associated effect on the synthesized graphene. They observed that the initial step needed for the process is solvent intercalation using electrical and in situ measurements. Secondly, how thick an exfoliated graphene is depends on an intercalation degree. They noted that with electrochemical disintegration of water, graphite tends to expand, thus controlling the lateral size of the exfoliated graphene.

11.5 MECHANICAL EXFOLIATION

Mechanical exfoliation involves splitting a layer of atoms into thin sheets. It is often regarded as one of the methods that involve exfoliation of bulky crystal structures. It can be conducted courtesy of electromagnetism in inert, vacuum, or air medium and also by using both electrostatic and mechanical means.

In the mechanical exfoliation process, one challenge that needs to be overcome is the Van der Waal force of attraction between graphene layers. Jayasena and Subbiah (2011) demonstrated a new approach to create a few graphene layers from bulk graphite through mechanical cleavage. Using ultrasonic oscillations – dependent extremely sharp single crystal diamond wedge, the authors were able to cleave a highly organized pyrolytic graphite specimen to form graphene layers. Products of this approach were characterized by graphene layers having areas of a few micrometers.

Ji et al. (2012) indicated that mixing graphite in appropriate stabilizing liquids leads to massive exfoliation that produces dispersed graphene nanosheets. Raman spectroscopy and X-ray photoelectron spectroscopy reveals that the exfoliated products are devoid of basal plane-based defects, and they are also unoxidized. The technique is appropriate for exfoliating molybdenum disulphide, boron nitride, and many other layered crystals.

11.6 LIQUID PHASE EXFOLIATION

Hernandez et al. (2008) examined dispersion of graphene with about 0.01mg/ml concentration created through exfoliation and dispersion of graphite in N-methyl-pyrrolidone. The authors discovered that using electron diffraction, Raman spectroscopy and electron microscopy, individual graphene sheets are present. With this method, approximately 1wt percent yield was produced, which is capable of increasing to 7–12wt percent if subjected to further processing.

11.7 PYROLYSIS

Decomposition of complex molecules using heat is called pyrolysis. Subrahmanyam et al. (2009) reported that graphene can be prepared through many methods, including pyrolysis of camphor in reductive conditions. The authors documented a direct chemical synthesis of carbon nanosheets using common reagents such as ethanol and sodium which, in a reaction produce an intermediate solid product that is then pyrolized to yield a jointed array of graphene sheets that can be dispersed by mild sonication.

11.8 REDUCTION OF GRAPHENE OXIDE

Chemical reduction of graphite oxide is one of the usual procedures for preparing graphene in a large scale. Hummer and Offeman (1958) reported that treatment with graphite with water free mix of sodium nitrate, potassium permanganate and sulfuric acid. Less than two hours is required for the entire process to complete at temperatures lower than 45 degrees Celsius Shin et al. (2009) reported the modulation of the conductivity of graphite oxide films with the help of reducing agents. It was discovered that the resistance of graphite oxide film decreased with the aid of borohydride than that of films reduced by hydrazine.

Zhou et al. (2011) documented a fast and cost-saving approach for reducing graphene oxide (GO) through hydroxylamine as a reductant. When this method was compared with other reported methods, graphene oxide reduction using hydroxylamine appears as a platform much more preferable to bulk production of the graphene owing to its simplicity, cost-effectiveness and efficiency.

Zhu et al. (2010) designed an easy approach that can be devised in the synthesis of chemically converted graphene nanosheets (GNS). This conversion was based on reducing sugars like fructose, glucose and sucrose with the aid of exfoliated graphite oxide (GO) as precursor. The advantage of this method is that the oxidized and reducing agents involved are environmentally friendly.

Zhang et al. (2010) showed that individual graphene oxide sheets can be readily reduced with the aid of L-ascorbic acid under mild conditions. In the study conducted by Wang et al. (2008), a large quantity of graphene nanosheets were synthesized through soft chemistry synthetic means. This involves ultrasonic exfoliation, graphite oxidation and chemical reduction. Graphene nanosheets with corrugations that look like ripples and having 10 to 100 square nanometers were synthesized through electron microscopy, X-ray diffraction and transmission electron microscopy (TEM) observations.

Fan et al. (2008) made stable graphene suspensions with the aid of novel and intriguing synthetic approaches involving fast deoxygenation of exfoliated graphite oxide in strong basic solutions at reduced temperatures between 50°C and 90°C and without reducing agents. This work highlights the possibility of producing graphene on an industrial scale.

Sundaram et al. (2008) reported that lead particles attachment to graphene renders it hydrogen sensitive. Comptom et al. (2011) reported that organic dispersions of graphene oxide can be reduced thermally in polar organic solvents in the presence of reflux conditions in order to install electrically conductive, chemically active reduced graphene oxide (CARGO) with adjustable carbon to oxygen ratios, which will be based on the solvent boiling point. Hydroxyl and carboxyl groups can be retrieved when reflux is conducted at temperature above 155 °C. Reductions can be attained only one hour post-reflux without corresponding carbon to oxygen ratios changing even in the subsequent treatment.

11.9 SONICATION

Paredes et al. reported that sonication can be used to fully exfoliate graphite oxide substance into individual single-layered graphene sheets. They noted the dispersed graphene oxide was consist with sheets ranging between hundred nanometers and few micrometers. Graphene-based devices with different applications can be manipulated by the results of the study.

Dubin et al. (2010) refluxed graphene oxide in N-methyl-2-pyrrolidinone (NMP) resulting in reduction and deoxygenation to produce stable colloidal dispersion. They reported that solvothermal reduction is followed by a change in color from light brown to black. Using and scanning electron microscopy and atomic force microscopy (SEM) images of the flakes, the presence of single sheets of the solvothermally decreased graphene oxide (SRGO) is confirmed.

11.10 CHEMICAL VAPOUR DEPOSITION

Maria del Prado et al. (2016) documented how to enhance both Hummers method and chemical vapor deposition. In the work, nickel and copper were utilized as catalyzed in the chemical vapor deposition. Hydrogen and methane were also used as precursor gases. The result showed that a peak thickness was produced for samples that were created at 1050°C.

Li et al. (2009) produced centimeter-dimension graphene films through chemical vapor deposition with the aid of methane. The films are majorly single-layered graphene with less than 5 percent having few layers. Owing to the low solubility of carbon in copper, the process of growing graphene films becomes self-limited.

Obratzsov et al. (2007) developed well-ordered graphene films having a thickness of a few graphene layers through chemical vapor deposition. The films were developed on nickel substrates. With respect to data from Raman microscopy, auger microscopy and scanning tunnel microscopy, the chemical vapor deposition film thickness is calculated as 1.5 ± 0.5nm. A smooth surface alongside with atomic arrangement in between ridges is shown.

Bae et al. (2010) documented wet-chemical and roll-to-roll synthesis of largely 76.2cm graphene produced by chemical vapor deposition into flexible copper substrates. They reported that the films have sheet resistance that is very low, which is higher in quality than indium tin oxides and other commercial electrodes. Layer-by-layer stacking was used in the formulation of a doped four-layered film.

Li et al. (2011) grew graphene single crystals measuring up to 0.5mm on a side through low pressure chemical vapor deposition in copper foil surroundings using a methane precursor. It was discovered that large graphene regions exhibited a crystallographic orientation using low-energy electron microscopy.

11.11 GRAPHENE FROM NATURAL SOURCES: SYNTHESIS, CHARACTERIZATION AND APPLICATIONS

Graphene and graphene-derived substances have sparked a positive response in the lab because of its unique physical, chemical, and physiochemical features, which also have positioned graphene as a promising resource for optics, energy-harvesting and future electronics devices. Graphene possesses outstanding mechanical characteristics and chemical inertness, as well as great mobility and optical transparency. Single-layer graphene has an extreme high optical relative permeability (98%) that permits light to travel through some kind of wide range of wavelengths, making it an ideal medium for an optically conductive window. Additionally, multiple chemical surface modification techniques may be used to control graphene's optical, electrocatalytic properties and electrical features, making it one of the best choices for sophisticated technologies in optoelectronic and energy-harvesting devices. This study outlines the most-significant experimental findings from the results of research on the intriguing features of graphene electrodes and their use in different types of solar cells. In addition, the state of the art of several graphene production techniques and surface modification for photovoltaic applications is covered in this chapter (Das et al., 2014).

Graphene has sparked a lot of interest in the science establishment since Novoselov, Geim, and co-workers demonstrated its facile isolation through exfoliating graphite in 2004. Due to its highly unique qualities, it has given great hopes for innovative applications in a wide range of fields. Analog electronics as well as optoelectronics/photonics are some of the potential applications that will be briefly discussed here. Of course, the graphite exfoliating as a resource of graphene for advanced materials in this research, so we start with large-scale graphene growth processes (Avouris and Dimitrakopoulos, 2012).

The area of nanotechnology has progressed since Geim and Novoselov, who in 2004 discovered graphene, a two-dimensional substance made up of sp2 hybridized carbon atoms. Because of its outstanding electrical, thermal, mechanical, and optical capabilities, as well as its huge surface area and single-atom thickness, graphene has received a lot of interest. As a result, numerous processes for obtaining graphene have been discovered, including chemical exfoliation, chemical vapor deposition (CVD), organic synthesis, and so on. Nevertheless, generating graphene with fewer flaws and at a large scale is a big difficulty for these procedures; hence, there is a rising demand to generate graphene in high amounts with good quality. Several investigations have been conducted in order to discover new ways to produce elevated graphene. This work is primarily concerned with the production of graphene and its precursor, graphene oxide. Characterization techniques to distinguish graphene include scanning electron microscopy (SEM), optical microscopy, scanning probe microscopy (SPM), Raman spectroscopy utilized to ascertain amount of quality, layers, structures, atomic and defects in graphene (Adetayo & Runsewe, 2019).

As the semiconductor industry approached the end of the accelerating Moore's roadmap for device downscaling in the previous decade, the need to identify new candidate materials drove numerous research studies to investigate a wide range of non-conventional substances. Graphene, carbon nanotubes, and organic conductors seem to be the most effective among them. To continue the ever-shrinking road of nanoelectronics, finding a material having metallic capabilities mixed with field effect characteristics on a tiny level will always be a dream. The atomically thin carbon layer, graphene, has attracted the industry's attention not only in micro-, nano-, and opto-electronics, but also in biotechnology, owing to its fantastic features such as optical transparency, high mobility, mechanical stiffness, room temperature quantum Hall effect, and so on. The material's compliance using Si processing technique is a critical feature for mass manufacture. Several graphene development processes will be reviewed and contrasted, as well as a wide range of graphene uses. Also, there is a brief assessment of the performance results of various devices. Studies have mostly focused on background physics and its use for electrical devices. However, according to recent research on its uses in photonics and optoelectronics, this substance can achieve ultra-wide band optimization with or without a bandgap, unique combination of its distinctive optical and electrical properties (Akbar et al., 2015).

The nanotechnology research team has a lot of expertise with anodized alumina as a substrate for nanostructures. We explored and created a platform for fabricating copper nanowires utilizing an anodized alumina template that can be used for graphene synthesis utilizing the plasma accelerated chemical deposition methodology in this study. The adhesive capabilities of these nanostructures on the substrates were also examined as part of the project. Furthermore, the discovery of a new methodology for fabricating nanoporous thin film membranes is an important feature of this effort. The features of graphene, one-atom-thick carbon sheets, have prompted academics and businesses to examine its manufacture, characteristics, and uses in a variety of sectors. Graphene of high grade is physically strong, light, almost transparent, and an excellent heat and electrical conductor. Its interconnections with many other substances and light, as well as its two-dimensional structure, give it special properties like the bipolar transistor impact, ballistic charge transfer, and massive quantum oscillations. The following overview illustrates graphene's various possible applications (Abbasi et al., 2014).

Over the last few years, graphene has piqued scientists' curiosity due to its outstanding electrical, mechanical, and chemical characteristics. Advanced electrical gadgets and composite materials should benefit from this strictly two-dimensional (2D) material. The issue is to generate vast areas of defect-free graphene, which is required for electronic applications, while bulk synthesis of graphene with flaws, which allows for nanoparticle attachment sites, is required for applications such as catalysis. The spatial characterization of the few layered transparent graphenes through micro Raman spectroscopy, its electrical characterization exhibiting p-semiconductor behavior, and research on the gas-sensing characteristics with small quantities of CO as well as H2 are all described in this chapter. Graphene development has also been carried out on Cu-catalyst surfaces using APCVD and low pressure CVD (LPCVD). Impact of metal catalyst thickness and CVD growth parameters (concentrations of the gases, growth period, evaporative cooling, etc.) were researched in detail to optimize the quality of graphene with respect to the number of layers and flaws.

Commercial application of graphene was established by oxidation of graphite to graphite oxide (GO) and accompanied through reduction reaction. Different methods for synthesizing GO (Staudenmaier's, Hummers', Modified Hummers', and Tour's Methods) were compared in order to obtain highly oxidized GO. To increase the quality of chemically generated graphene, various reduction procedures were investigated (CDG). Pd/CDG was used as a catalyst in the dehydrogenation and hydrolysis of ammonia borane, and metal oxide nanoparticles (NPs of Ni, Au, as well as Pd) were successfully supported on CDG (AB). In addition, titanium dioxide (by UV-assisted and hydrothermal methods), tungsten oxide (by sonochemical method), and zinc oxide (through thermal degradation technique) metal oxide NPs were systematically investigated on CDG. Under UV light, CDG composites with WO_3 and TiO_2 were successfully used in photodegradation reactions of methylene blue (MB). At ambient temperature and at high temperatures, sensitivity investigations of CDG/ZnO synthetic polymers were carried out towards hydrogen gas (Trivedi et al., 2019).

Graphene oxide (GO) and reduced graphene oxide (rGO) have created new possibilities for gas barrier, ion exchange, and stimuli-response features in nanocomposites due to their facile top-down manufacturing. We examine the most prevalent synthesis procedures for producing those graphene derivatives, describe how production influences their main mechanical properties, and show several examples of nanocomposites with unique and amazing features in this chapter. Extraction techniques, stimuli-responsive composites, anti-corrosion coatings, as well as energy storage are all highlighted in the chapter. Finally, we examine the future of industrial-scale manufacture and application of graphene-derived polymer nanocomposites, as well as the outstanding hurdles (Andrew et al., 2019).

11.12 CONCLUSION

This chapter has provided comprehensive details on the application, characterization and synthesis of graphene. Due to ever-increasing demands for a sustainable future, research focus on the improvement in the quality of graphene production through a green approach will ensure emergence of top-notch, non-toxic, sustainable graphene and enhance the market value. The unique features and properties of graphene have provided an opportunity for fabricating integrated heterostructures materials for diverse applications (Adetunji et al., 2021a,b,c; Ukhurebor et al., 2021; Sangeetha et al., 2021; Adetunji and Ukhurebor, 2021, Oluwaseun et al., 2017; Adetunji et al., 2012; Arowora et al., 2012; Dauda et al., 2022a,b; Okeke et al., 2021; Adetunji and Anani, 2021; Nwankwo et al., 2021; Adejumo and Adetunji. 2018; Ukhurebor et al 2021).

REFERENCES

Abbasi, E., Akbarzadeh, A., Kouhi, M., & Milani, M. (2014). Graphene: Synthesis, bio-applications, and properties. *Artificial Cells, Nanomedicine, and Biotechnology*, 44(1), 150–156. Doi:10.3109/21691401.2014.927880

Achee, T.C., Sun, W., Hope, J.T., et al. (2018). High-yield scalable graphene nanosheet production from compressed graphite using electrochemical exfoliation. *Science Report*, 8, 14525. https://doi.org/10.1038/s41598-018-32741-3

Adejumo, I.O., & Adetunji. C.O. (2018). Production and evaluation of biodegraded feather meal using immobilised and crude enzyme from *Bacillus subtilis* on broiler chickens. *Brazilian Journal of Biological Sciences*, 5(10), 405–416.

Adetayo, A., & Runsewe, D. (2019). Synthesis and fabrication of graphene and graphene oxide: A review. *Open Journal of Composite Materials*, 9, 207–229. https://doi.org/10.4236/ojcm.2019.92012

Adetunji, C.O., & Anani, O.A. (2021). Bioaugmentation: A powerful biotechnological techniques for sustainable ecorestoration of soil and groundwater contaminants. In: Panpatte, D.G., & Jhala, Y.K. (eds.), *Microbial Rejuvenation of Polluted Environment. Microorganisms for Sustainability*, vol. 25. Springer: Singapore. https://doi.org/10.1007/978-981-15-7447-4_15

Adetunji, C.O., Anani, O.A., Olaniyan, O.T., Inobeme, A., Olisaka, F.N., Uwadiae, E.O., & Obayagbona, O.N. (2021a). Recent trends in organic farming. In: Soni, R., Suyal, D.C., Bhargava, P., & Goel, R. (eds.), *Microbiological Activity for Soil and Plant Health Management*. Springer: Singapore. https://doi.org/10.1007/978-981-16-2922-8_20

Adetunji, C.O., Fawole, O.B., Afolayan, S.S., Olaleye, O.O., & Adetunji, J.B. (2012). An Oral Presentation During 3rd NISFT Western Chapter Half Year Conference/General Meeting, Ilorin, pp. 14–16.

Adetunji, C.O., Jeevanandam, J., Anani, O.A., Inobeme, A., Thangadurai, D., Islam, S., & Olaniyan, O.T. (2021b). Microbially-derived biosurfactants for improving sustainability in industry. In: *Green Sustainable Process for Chemical and Environmental Engineering and Science*, pp. 299–315. https://doi.org/10.1016/B978-0-12-823380-1.00002-2

Adetunji, C.O., Michael, O.S., Nwankwo, W., Ukhurebor, K.E., Anani, O.A., Oloke, J.K., Varma, A., Kadiri, O., Jain, A., & Adetunji, J.B. (2021c). Quinoa, the next biotech plant: Food security and environmental and health hot spots. In: Varma, A. (eds.), *Biology and Biotechnology of Quinoa*. Springer: Singapore. https://doi.org/10.1007/978-981-16-3832-9_19

Adetunji, C.O., & Ukhurebor, K.E. (2021). Recent trends in utilization of biotechnological tools for environmental sustainability. In: Adetunji, C.O., Panpatte, D.G., & Jhala, Y.K. (eds.), *Microbial Rejuvenation of Polluted Environment: Microorganisms for Sustainability*, vol. 27. Springer: Singapore. https://doi.org/10.1007/978-981-15-7459-7_11

Akbar, F., Kolahdouz, M., Larimian, S., Radfar, B., & Radamson, H.H. (2015). Graphene synthesis, characterization and its applications in nanophotonics, nanoelectronics, and nanosensing. *Journal of Materials Science: Materials in Electronics*, 26(7), 4347–4379. Doi:10.1007/s10854-015-2725-9

Andrew, T.S., LaChance, A.M., Zeng, S., Liu, B., & Luyi, S. (2019). Synthesis, properties, and applications of graphene oxide/reduced graphene oxide and their nanocomposites, *Nano Materials Sciences*, 1(1), 31–47. https://doi.org/10.1016/j.nanoms.2019.02.004.

Arowora, K.A., Abiodun, A.A., Adetunji, C.O., Sanu, F.T., Afolayan, S.S., & Ogundele, B.A. (2012). Levels of aflatoxins in some agricultural commodities sold at Baboko Market in Ilorin, Nigeria. *Global Journal of Science Frontier Research*, 12(10), 31–33.

Avouris, P., & Dimitrakopoulos, C. (2012). Graphene: Synthesis and applications. *Materials Today*, 15(3), 86–97. Doi:10.1016/s1369-7021(12)70044-5.

Bae, S., Kim, H., Lee, Y., et al. (2010). Roll-to-roll production of 30-inch graphene films for transparent electrodes. *Nature Nanotechnology*, 5(8), 574–578. Doi:10.1038/nnano.2010.132

Bhuyan, S.A., Uddin, N., Islam, M., Bipasha, F.A., & Hossain, S.S. (2016). Synthesis of graphene. *International Nano Letters*, 6, 65–83. Doi:10.1007/s40089-015-0176-1

Compton, O.C., Jain, B., Dikin, D.A., Abouimrane, A., Amine, K., Nguyen, S.T. (2011). Chemically active reduced graphene oxide with tunable C/O ratios. *ACS Nano*, 5(6), 4380–4391 Doi:10.1021/nn1030725

Das, S., Sudhagar, P., Kang, Y., & Choi, W. (2014). Graphene synthesis and application for solar cells. *Journal of Materials Research*, 29(3), 299–319. Doi:10.1557/jmr.2013.297

Dauda, W.P., Abraham, P., Glen, E., Adetunji, C.O., Ghazanfar, S., Ali, S., Al-Zahrani, M., Azameti, M.K., Alao, S.E.L., Zarafi, A.B., Abraham, M.P., & Musa, H. (2022a). Robust profiling of cytochrome P450s (P450ome) in notable *Aspergillus* spp. *Life*, 12(3), 451. https://doi.org/10.3390/life12030451

Dauda, W.P., Morumda, D., Abraham, P., Adetunji, C.O., Ghazanfar, S., Glen, E., Abraham, S.E., Peter, G.W., Ogra, I.O., Ifeanyi, U.J., Musa, H., Azameti, M.K., Paray, B.A., & Gulnaz, A. (2022b). Genome-wide analysis of cytochrome P450s of *Alternaria* species: Evolutionary origin, family expansion and putative functions. *Journal of Fungi*, 8(4), 324. https://doi.org/10.3390/jof8040324

Dubin, S., Gilje, S., Wang, K., Tung, V.C., Cha, K., Hall, A.S., Farrar, J., Varshneya, R., Yang, Y., & Kaner, R.B. (2010). A one-step, solvothermal reduction method for producing reduced graphene. *ACS Nano*, 4(7), 3845–3852. Doi:10.1021/nn100511a

Fan, X., Peng, W., Li, Y., Li, X., Wang, S., Zhang, G., & Zhang, F. (2008). Deoxygenation of exfoliated graphite oxide under alkaline conditions: A green route to graphene preparation. *Advanced Material*, 20(23), 4490–4493. Doi:10.1002/adma.200801306

Hernandez, Y., Nicolosi, V., Lotya, M., Blighe, F. M., Sun, Z., De, S., McGovern, I.T., Holland, B., Byrne, M., Gun'Ko, Y.K., Boland, J.J., Niraj, P., Duesberg, G., Krishnamurthy, S., Goodhue, R., Hutchison, J., Scardaci, V., Ferrari, A.C., & Coleman, J. N. (2008). High-yield production of graphene by liquid-phase exfoliation of graphite. *Nature Nanotechnology*, 3(9), 563–568. Doi:10.1038/nnano.2008.215. Epub 2008 Aug 10. PMID: 18772919

Hofmann, M., Chiang, W.Y., Nguyễn, T.D., & Hsieh, Y.P. (2015). Controlling the properties of graphene produced by electrochemical exfoliation. *Nanotechnology*, 26(33), 335607. Doi:10.1088/0957-4484/26/33/335607. Epub 2015 Jul 29. PMID: 26221914

Hummers, W.S., & Offeman, R.E. (1958). Preparation of graphitic oxide. *Journal of the American Chemical Society*, 80(6), 1339. Doi:10.1021/ ja01539a017

Jayasena, B., & Subbiah, S. (2011). A novel mechanical cleavage method for synthesizing few-layer graphenes. *Nanoscale Research Letters*, 6(1), 95. Doi:10.1186/1556-276X-6-95 PMID: 21711598; PMCID: PMC3212245

Ji, L, Xin, H.L., Kuykendall, T.R., Wu, S.L., Zheng, H., Rao, M., Cairns, E.J., Battaglia, V., & Zhang, Y. (2012). SnS2 nanoparticle loaded graphene nanocomposites for superior energy storage. *Physical Chemistry Chemical Physics Journal*, 14(19), 6981–6986. Doi:10.1039/c2cp40790f. Epub 2012 Apr 12. PMID: 22495542

Jiang, J., Chen, T., Xia, Y., Jia, Z., Liu, Z., & Zhang, H. (2014). Synthesis, characterization, and tribological behavior of oleic acid capped graphene oxide. *Journal of Nanomaterials*, 2014, 654145. Doi:10.1155/2014/654145

Lavín López, P., Valverde Palomino, J.L., Sánchez Silva, M.L., & Izquierdo, A.R. (2016). Optimization of the synthesis procedures of graphene and graphite oxide. In: Nayak, P.K. (ed.), *Recent Advances in Graphene Research.* IntechOpen (October 12). Doi:10.5772/63752. Available from: www.intechopen.com/books/recent-advances-in-graphene-research/optimization-of-the-synthesis-procedures-of-graphene-and-graphite-oxide

Li, X., Magnuson, C.W., Venugopal, A., Tromp, R.M., Hannon, J.B., Vogel, E.M., Colombo, L., & Ruoff, R.S. (2011). Large-area graphene single crystals grown by low-pressure chemical vapor deposition of methane on copper. *Journal of the American Chemical Society*, 133, 2816. Doi:10.1021/ja109793s

Li, X.S., Cai, W.W., An, J.H., Kim, S., Nah, J., Yang, D.X., Piner, R., Velamakanni, A., Jung, I., Tutuc, E., Banerjee, S.K., Colombo, L., & Ruoff, R.S. (2009). Large-area synthesis of high-quality and uniform graphene films on copper foils. *Science*, 324(5932), 1312–1314. Doi:10.1126/science.1171245

Liu, F., Wang, C., Sui, X., Riaz, M.A., Xu, M., Wei, L., & Chen, Y. (2019). Synthesis of graphene materials by electrochemical exfoliation: Recent progress and future potential. *Carbon Energy*, 1(2), P173–199. Doi:10.1002/cey2.14

Mao, H.Y., Lu, Y.H., Lin, J.D., Zhong, S., ShenWee, A.T., & Chen, W. (2013). Manipulating the electronic and chemical properties of graphene via molecular functionalization. *Progress in Surface Science*, 88(2), 132–159.

McAllister, M.J., Li, J.L., Adamson, D.H., Schniepp, H.C., Abdala, A.A., Liu, J., Alonso, M.H., Milius, D.L., Car, R., Robert, K., Prud'homme, R.K., & Aksay, I.A. (2007). Single sheet functionalized graphene by oxidation and thermal expansion of graphite. *Chemistry of Materials*, 19(18), 4396–4404. Doi:10.1021/ cm0630800

Nwankwo, W., Adetunji, C.O., Ukhurebor, K.E., Panpatte, D.G., Makinde, A.S., & Hefft, D.I. (2021). Recent advances in application of microbial enzymes for biodegradation of waste and hazardous waste material. In: Adetunji, C.O., Panpatte, D.G., & Jhala, Y.K. (eds.), *Microbial Rejuvenation of Polluted Environment: Microorganisms for Sustainability*, vol 27. Springer: Singapore. https://doi.org/10.1007/978-981-15-7459-7_3

Obraztsov, A.N., Obraztsova, E.A., Tyurnina, A.V., & Zolotukhin, A.A. (2007). Chemical vapor deposition of thin graphite films of nanometer thickness. *Carbon*, 45(10), 2017–2021. Doi:10. 1016/j.carbon.2007.05.028

Okeke, N.E., Adetunji, C.O., Nwankwo, W., Ukhurebor, K.E., Makinde, A.S., & Panpatte, D.G. (2021). A critical review of microbial transport in effluent waste and sewage sludge treatment. In: Adetunji, C.O., Panpatte, D.G., & Jhala, Y.K. (eds.), *Microbial Rejuvenation of Polluted Environment: Microorganisms for Sustainability*, vol. 27. Springer: Singapore. https://doi.org/10.1007/978-981-15-7459-7_10

Oluwaseun, A.C., Phazang, P., & Sarin, N.B. (2017). Significance of rhamnolipids as a biological control agent in the management of crops/plant pathogens. *Current Trends in Biomedical Engineering & Biosciences*, 10(3), pp. 54–55.

Sahu, D., Sutar, H., Senapati, P., Murmu, R., & Roy, D. (2021) Graphene, graphene-derivatives and composites: Fundamentals, synthesis approaches to applications. *Journal of Composites Science*, 5, 181. https://doi.org/10.3390/jcs5070181

Sangeetha, J., Hospet, R., Thangadurai, D., Adetunji, C.O., Islam, S., Pujari, N., & Al-Tawaha, A.R.M.S. (2021). Nanopesticides, nanoherbicides, and nanofertilizers: The greener aspects of agrochemical synthesis using nanotools and nanoprocesses toward sustainable agriculture. In: Kharissova, O.V., Torres-Martínez, L.M., & Kharisov, B.I. (eds.), *Handbook of Nanomaterials and Nanocomposites for Energy and Environmental Applications*. Springer: Cham. https://doi.org/10.1007/978-3-030-36268-3_44

Shin, H.-J., Kim, K.K., Benayad, A., Yoon, S.-M., Park, H.K., Jung, I.-S., Jin, M.H., Jeong, H.-K., Kim, J.M., Choi, J.-Y., & Lee, Y.H. (2009). Efficient reduction of graphite oxide by sodium borohydride and its effect on electrical conductance. *Advanced Functional Materials*, 19(12), 1987–1992. Doi:10.1002/adfm.200900167 103

Subrahmanyam, K.S., Vivekchand, S.R.C., Govindaraj, A., & Rao, C.N.R. (2008). A study of graphenes prepared by different methods: Characterization, properties and solubilization, *Journal of Materials Chemistry*, 18(13), 1517–1523. http://dx.doi.org/10.1039/B716536F

Sundaram, R.S., G´omez-Navarro, C., Balasubramanian, K., Burghard, M., & Kern, K. (2008). Electrochemical modification of graphene. *Advanced Materials*, 20(16), 3050–3053. Doi:10.1002/ adma.200800198

Trivedi, S., Lobo, K., & Ramakrishna Matte, H.S.S. (2019). Synthesis, properties, and applications of graphene. *Fundamentals and Sensing Applications of 2D Materials,* 25–90. Doi:10.1016/b978-0-08-102577-2.00003-8

Ukhurebor, K.E., Adetunji, C.O., Bobadoye, A.O., Aigbe, U.O., Onyancha, R.B., Siloko, I.U., Emegha, J.O., Okocha, G.O., & Abiodun, I.C. (2021). Bionanomaterials for biosensor technology. *Bionanomaterials: Fundamentals and Biomedical Applications*, 5–22.

Ukhurebor, K.E., Mishra, P., Mishra, R.R., & Adetunji, C.O. (2020). Nexus between climate change and food innovation technology: Recent advances. In: Mishra, P., Mishra, R.R., & Adetunji, C.O. (eds.), *Innovations in Food Technology*. Springer: Singapore. https://doi.org/10.1007/978-981-15-6121-4_20

Ukhurebor, K.E., Nwankwo, W., Adetunji, C.O., & Makinde, A.S. (2021). Artificial intelligence and internet of things in instrumentation and control in waste biodegradation plants: Recent developments. In: Adetunji, C.O., Panpatte, D.G., & Jhala, Y.K. (eds.), *Microbial Rejuvenation of Polluted Environment. Microorganisms for Sustainability*, vol 27. Springer: Singapore. https://doi.org/10.1007/978-981-15-7459-7_12

Wang, G., Yang, J., Park, J., Gou, X., Wang, B., Liu, H., & Yao, J. (2008). Facile synthesis and characterization of graphene nanosheets. *Journal of Physical Chemistry*, 112(22), 8192–8195. Doi:10.1021/ jp710931h

Wang, S, Ang, P.K., Wang, Z., Tang, A.L.L., Thong, J.T.L., & Loh, K.P. (2010). High mobility, printable, and solution-processed graphene electronics. *Nano Letters*, 10, 92.

Zhang, J., Yang, H., Shen, G., Cheng, P., Zhang, J., & Guo, S. (2010). Reduction of graphene oxide via L-ascorbic acid. *Chemical Communications*, 46, 1112–1114. Doi:10.1039/B917705A.

Zhou, X., Zhang, J., Wu, H., Yang, H., Zhang, J., & Guo, S. (2011). Reducing graphene oxide via hydroxylamine: A simple and efficient route to graphene. *Journal of Physical Chemistry C*, 115(24), 11957–11961. Doi:10.1021/jp202575j

Zhu, C., Guo, S., Fang, Y., & Dong, S. (2010). Reducing sugar: New functional molecules for the green synthesis of graphene nanosheets. *ACS Nano*, 4(4), 2429–2437. Doi:10.1021/nn1002387

12 Artificial Intelligence in the Food Packaging Industry

Ogundolie Frank Abimbola and
Michael O. Okpara

CONTENTS

12.1 INTRODUCTION

The food packaging industry is steadily growing and evolving because of its importance in the food-supply chain. There is practically no food product in the market that does not require one form or another of food packaging because of the roles packaging plays in the food supply chain. Food packaging protects food items from pollutants and therefore prevents their spoilage during distribution and storage. Liquid and semi-solid food items are prevented from spillage through the protective function of food packaging. Moreover, for many customers, food packaging makes food items more appealing and increases the chances of improving sales.

The history of food packaging is as old as ancient Egyptian civilization and has evolved especially with the advent of technology. The science of food packaging is believed to have started with the storage of food in glass containers in ancient Egypt (Grayhurst & Girling, 2011). Over the centuries, food packaging has seen some advancements with the use of wooden boxes, ceramics, fabrics, paper, tin and aluminum cans, plastic bottles, antimicrobial films, and so forth. Today the application of artificial intelligence (AI) in food packaging makes it possible to sort food items for packaging, monitor their transportation, manage food inventory in warehouses and stores, track food fraud and theft, authenticate food items, provide nutrition information about the food item, detect food spoilage, retrieve spoilt food, and hasten check-out process in stores.

DOI: 10.1201/9781003207955-12

12.2 HISTORY OF AI IN FOOD PACKAGING

Food packaging is an essential component in food production, processing, safety, distribution and freshness from the farm, or from the factory of the producer, to the plate of the consumer. Over the centuries, there has been a continuous and steady evolution in food packaging, and this has ensured that a very large proportion of food materials have been prevented from spoilage and wastage. The history of food packaging can be traced back five thousand years, when glass was used to store food in Egypt (Grayhurst & Girling, 2011).

Subsequently, other materials like wood, ceramics, and fabrics were used as alternatives for food packaging. The use of paper for food packaging emerged in the second century, when paper was invented by Emperor Ts'ai Lun in China (Li, 1974). The flexibility of paper made it easily acceptable for food packaging, as paper can easily be folded into many different shapes for packaging. The use of paper for food packaging continued until today, with many innovations to its application. From 500 to 1500 AD, the use by voyagers of wooden boxes and barrels for food packaging/storage became prominent. The long time they spent on their voyages necessitated devising larger food-packaging material. The use of canning for food packaging came into the limelight in the late 1790s, when Nicholas Appert introduced the use of corked glass containers for packaging and preservation (Featherstone, 2012; Graham, 1981). Appert's canning technique was applied by the French army to package and preserve their food during the Napoleonic Wars.

In 1810, an innovation to the canning technique was made by Peter Durand, who used tin instead of glass to make cans for food packaging and preservation. Durand patented his innovation in 1810 and by 1813, his innovation was commercially applied in supplying canned food to the Royal Navy (Featherstone, 2012). The use of tin cans for food packaging and preservation is still in application today. The use of aluminum foil containers for food packaging came into prominence in 1910; and in the late 1950s, aluminum cans were produced for food packaging, pioneered by Coors Brewing Company. Today, many food manufacturing companies use aluminum cans for food packaging. In 1973, Nathaniel Wyeth (an American engineer and inventor) used polyethylene terephthalate to produce plastic bottles for packaging carbonated drinks. And in 1980, the use of plastic material for packaging semi-solid and solid food materials came into prominence. Until today, plastic materials are still in use for packaging various kinds of drinks and food.

From the early 2000s to the present day, innovations in food packaging, targeted at elongating the shelf life of the food material, have been achieved. Antimicrobial film packaging bags can be used to prevent microbial growths and cross-contamination in packaged food before consumption. There are also silica gel packets that are efficient in absorbing moisture that can potentially initiate food spoilage. Other innovations that are presently applied in food packaging is the use of AI-based data carrier technologies like barcodes, quick response (QR) codes, and radio frequency identification (RFID) for the storage and transmission of important information about packaged food items.

Barcodes are machine-readable codes that consist of bars and spaces of varying width accompanied with some digits and are used for entering product information into a computer. Barcodes were invented and patented in 1952 by Norman Woodland and Bernard Silver, who designed vertical and circular barcoding systems (US12241649A, 1949). However, it was only in 1974 that George Laurer's vertical barcoding system was first used for commercial food packaging at a grocery store in Ohio (Ayres, 2021). The barcoding system is an example of an AI-based data carrier technology that can be used to store important product information like price, manufacture date, expiry date, batch number, and nutritional content of the packaged food. The stored product information on the barcodes can then be read with the aid of a barcode scanner. Barcodes are designed to code information in one dimension in a simple form consisting of parallel vertical bars and spaces that code for the product information.

Barcodes are now an important component in food packaging and are commonly used in retailing of food items among other commodities in supermarkets. Barcodes in food packaging ensure proper tracking of the items, hasten the check-out process, and reduces theft. When compared to other AI-based data carrier technologies, barcodes are quite limited in the volume of product information they can store. Consequently, more sophisticated AI-based data carrier technology, like QR codes and RFID, are gaining prominence in food packaging and distribution.

The QR coding system was invented in 1994 by Masahiro Hara from Denso Wave (Denso Wave, n.d.). The QR coding system was designed to code for information in two dimensions that consist of dots, hexagons, and square and rectangular bars on a white background. The coded information on QR codes are read across and up or down. Compared to the barcoding system, the QR coding system can store more product information and is faster in reading that information, especially for customers. Not long after its design, the QR coding system was adopted by the food industry for its packaging. Scanning food QR codes can provide the customer with vital information about the food, such as manufacture and expiry dates, nutritional content and recipe, allergy information, and so on. QR codes can also be used by food suppliers and retailers to manage their inventory while also monitoring food items in transit to their point of sale.

RFID is another AI-based data carrier technology that has gained wide acceptance in food packaging. RFID tags (also called labels or transponders) contain microchips and are designed to emit radio waves that are directed to an antenna and translated into important readable data on a computer with an updated database (López-Gómez et al., 2015). The packaged food items are assigned unique electronic product codes (EPC), which makes it possible to identify an individual item from the others. RFID is very useful for taking stock of food items carrying the RFID tags, reducing wastage of food items, improving food traceability, authenticating food items, detecting counterfeits, and quickly retrieving damaged/spoilt product. One of the drawbacks of the RFID tagging system in food packaging is the difficulty in reading RFID tags on packages containing much liquid or metals. Liquids are known to absorb radio waves while metals can reflect them. Consequently, it becomes quite challenging to read the RFID tags on such food items. Another drawback of the RFID tagging system is the high cost of installing the system, especially in retail outlets. Moreover, RFID tags are relatively more expensive compared to other technologies like barcode tags, which perform almost similar functions in food packaging (López-Gómez et al., 2015).

Today, food packaging has been taken to a whole new height with advancements in AI. Machines/robots for food packaging have been designed, and technologies that monitor the state of the freshness of packaged food are now available. Food packaging also involves the sorting of food items to ensure that they have uniform volume, size, or color and that they are in a good state before they are packaged for distribution. Ideally, a visual inspection by humans may be sufficient for food sorting but would require more manpower at a relatively lower speed and higher cost for food sorting compared to an AI-assisted automated sorting system. Consequently, from 2010 to 2016, a team of researchers at the University of Nebraska – Lincoln, led by Tom Duckett, designed a user-trainable software technology called Trainable Anomaly Detection and Diagnosis (TADD).

TADD is a system that utilizes machine-learning algorithms and computer vision to specifically sort food items based on important physical properties. This AI system has been successfully used to sort blemished potatoes from non-blemished potatoes based on texture and color, in 2D with at least 89 percent success rate, and its application is not limited to potatoes only (Barnes, Duckett, Cielniak, Stroud, & Harper, 2010; Duckett et al., 2014). Further work by this group in collaboration with industrial collaborators improved the AI system's food-sorting approach beyond texture and color in 2D to 3D imaging and sorting based on the shape of food items. This AI system has improved the quality and safety of food packaged for distribution, minimized food waste, and improved food production efficiency (Duckett et al., 2014).

12.2.1 Application of AI in Food Packaging

The use of machine intelligence (Artificial intelligence) in the food industries over the years has led to improvement in production output and reduction in food safety concerns, thereby resulting in increased confidence in the packaged foods by enhancing food security (Singh and Bagade, 2022). It is achieved using computational applications. This productively allows optimization of major food problems in the world today, computerizing the various sectors of the food industry, ranging from manufacturing, processing, packaging and transportation, and also allows for the remodeling of products coming from various food and beverage industries (Soltani-Fesaghandis and Pooya, 2018; Kumar et al., 2021; Mavani et al., 2022). Automated food packaging has now also reduced workplace hazards that have been on the rise in the past decades.

Today artificial intelligence through smart or intelligent packaging has been able to increase the traceability of packaged food, thereby encouraging food monitoring and reducing fraud. The use of artificial intelligence has been able to increase the agility of food processing by using robots that require less coding but more machine learning. This technology of smart packaging has been applied in various forms of food packaging. e-Silva et al., (2021) reported the use of AI for the improvement of food packaging films from fish gelatin using different essential oils such as oregano oil, and oil palm to further increase the elasticity and antimicrobial properties of the packaging films.

In the various food industries requiring food packaging, ranging from beverage, confectioneries, bread, biscuits among others, the use of artificial intelligence today is engaged in the production of user friendly packages and also a food chain with effective packaging mechanism that is time-efficient and requires less manpower. Food industries now utilize AI for food security management through the production of uncloneable food package labels using barcodes, RFID and other forms of labels (Yam et al., 2005; Liu et al., 2019), and production of unique packaging labels centered on enticing the end consumer. These packages can also communicate by giving real-time information about the state of the product. The product information and package specificity can be achieved through the use of machine intelligence.

Another interesting area of the application of artificial intelligence in food packaging is in food sorting. The process of sorting of qualified and finished products can be painstaking and challenging. This process often requires multiple manpower and is time consuming. The use of big data, machine learning, deep machine learning and AI have give production companies an easy way out with AI-based systems making decisions based on fruits' texture, color, and shapes, among other parameters (Tripathi et al., 2020; Dewi et al., 2020; Kumar et al., 2021).

12.2.2 AI Systems Applied in Food Packaging

Digital transformation in the area of food packaging has increased the production outputs of various food packaging sectors. Today, an integral path of our life is the AI, as we utilize the artificial intelligence applications in our everyday activities. Various AI systems such as virtual assistants (VA), robots, Self-Optimising-Clean-In-Place (SOCIP) among others are now being utilized in food processing industries, especially food packaging.

The use of virtual assistance in food packaging has increased engagement between producers and consumers, which has further strengthened the confidence of consumers in some packaged foods, increased producer revenue through improved local, national and international sales Chowdhury and Morey, (2019), and stimulated discussion of the application of VA in packaging farm products and how it could help consumers in knowing basic information about the packaged foods.

The use of robotics in food industries started with the packaging sector, though it was not generally accepted initially, but improvements and advantages observed, which include reduced factory accidents of personnel and improved time efficiency have today increased the acceptability and application of this technology in food packaging (Iqbal et al., 2017). Today, the use of AI with robotics has further reduced the chances of errors experienced in these industries when it comes to food packaging (Mahalik and Nambiar, 2010).

Today the use of computer vision and robots have accelerated tremendous growths not only in the food sector, but also in several manufacturing sectors of the world economy, as this has found value in various industries, such as agriculture and medicine. Today, in the meat packaging industries, use of robots which are easy to sterilize, has helped reduced contamination (Holmes and Holcombe, 2010). Robots are also used in picking and packaging foods, and they are automated to ensure seamless production and delivery of packaging. Depending on the programing, the use of automation can be adapted to various packaging stages.

Turnover period and sterilization of machines in the food production and packaging line has been boosted with the emergence of Self-Optimising-Clean-In-Place (SOCIP). This ensures that the surfaces of the machines are sterilized and cleaned without the need for the time consuming activities otherwise required to disassemble and assemble the machines.

Artificial intelligence has been an important improvement in the area of food packaging, which has not only improved production, but has also improved the confidence of consumers, and built trust in the manufacturing process. This application has increased the output of food production and, over the years, there has been a reduction in the record of casualties occurring in the packaging section of the various food industries.

REFERENCES

Ayres, R. U. (2021). Machine computation and digitization. In R. U. Ayres (Ed.), *The History and Future of Technology* (pp. 469–517). Cham: Springer, Cham. https://doi.org/10.1007/978-3-030-71393-5

Barnes, M., Duckett, T., Cielniak, G., Stroud, G., & Harper, G. (2010). Visual detection of blemishes in potatoes using minimalist boosted classifiers. *Journal of Food Engineering*, 98(3), 339–346. https://doi.org/10.1016/j.jfoodeng.2010.01.010

Chowdhury, E. U., & Morey, A. (2019). Intelligent packaging for poultry industry. *Journal of Applied Poultry Research*, 28(4), 791–800.

Denso Wave. (n.d.). QR code development. Retrieved March 16, 2022, from www.denso-wave.com/ja/technology/vol1.html

Dewi, T., Risma, P., & Oktarina, Y. (2020). Fruit sorting robot based on color and size for an agricultural product packaging system. *Bulletin of Electrical Engineering and Informatics*, 9(4), 1438–1445.

Duckett, T., Barnes, M., Hutton, J., Cielniak, G., Harper, G., Stroud, G., … Malekmohamadi, H. (2014). User-trainable visual anomaly detection for quality inspection tasks in the food industry. Retrieved March 16, 2022, from https://impact.ref.ac.uk/casestudies/CaseStudy.aspx?Id=40373

e-Silva, N. D. S., de Souza Farias, F., dos Santos Freitas, M. M., Hernández, E. J. G. P., Dantas, V. V., Oliveira, M. E. C., Joele, M. R. S. P., & Lourenço, L. D. F. H., (2021). Artificial intelligence application for classification and selection of fish gelatin packaging film produced with incorporation of palm oil and plant essential oils. *Food Packaging and Shelf Life*, 27, 100611.

Featherstone, S. (2012). A review of development in and challenges of thermal processing over the past 200years – A tribute to Nicolas Appert. *Food Research International*, 47(2), 156–160. https://doi.org/10.1016/j.foodres.2011.04.034

Graham, J. C. (1981). The French connection in the early history of canning. *Journal of the Royal Society of Medicine*, 74(5), 374–381. https://doi.org/10.1177/01410768810 7400511

Grayhurst, P., & Girling, P. J. (2011). Packaging of food in glass containers. In R. Coles & M. Kirwan (Eds.), *Food and Beverage Packaging Technology* (2nd ed., pp. 137–156). Blackwell.

Holmes, J. F., & Holcombe, W. D. (2010). Guidelines for designing washdown robots for mea packaging applications. *Trends in Food Science & Technology*, 21(3), 158–163.

Iqbal, J., Khan, Z. H., & Khalid, A. (2017). Prospects of robotics in food industry. *Food Science and Technology*, 37, 159–165.

Kumar, I., Rawat, J., Mohd, N., & Husain, S. (2021). Opportunities of artificial intelligence and machine learning in the food industry. *Journal of Food Quality*. https://doi.org/10.1155/2021/4535567

Li, H. L. (1974). An archaeological and historical account of Cannabis in China. *Economic Botany*, 28(4), 437–448.

Liu, Y., Han, F., Li, F., Zhao, Y., Chen, M., Xu, Z., Zheng, X., Hu, H., Yao, J., Guo, T., & Lin, W. (2019). Inkjet-printed unclonable quantum dot fluorescent anti-counterfeiting labels with artificial intelligence authentication. *Nature Communications*, 10(1), 1–9.

López-Gómez, A., Cerdán-Cartagena, F., Suardíaz-Muro, J., Boluda-Aguilar, M., Hernández-Hernández, M. E., López-Serrano, M. A., & López-Coronado, J. (2015). Radiofrequency identification and surface acoustic wave technologies for developing the food intelligent packaging concept. *Food Engineering Reviews*, 7(1), 11–32. https://doi.org/10.1007/s12393-014-9102-y

Mahalik, N. P., & Nambiar, A. N. (2010). Trends in food packaging and manufacturing systems and technology. *Trends in Food Science & Technology*, 21(3), 117–128.

Mavani, N. R., Ali, J. M., Othman, S., Hussain, M. A., Hashim, H., & Rahman, N. A. (2022). Application of artificial intelligence in food industry – a guideline. *Food Engineering Reviews*, 14(1), 134–175. https://doi.org/10.1007/s12393-021-09290-z

Singh, N., & Bagade, P. (2022). Quality, food safety and role of technology in food industry. *Thermal Food Engineering Operations*, 415–454. https://doi.org/10.1002/9781119776437.ch14

Soltani-Fesaghandis, G., & Pooya, A. (2018). Design of an artificial intelligence system for predicting success of new product development and selecting proper market-product strategy in the food industry. *International Food and Agribusiness Management Review*, 21(7), 847–864.

Tripathi, S., Shukla, S., Attrey, S., Agrawal, A., & Bhadoria, V. S. (2020). Smart industrial packaging and sorting system. In *Strategic System Assurance and Business Analytics* (pp. 245–254). Singapore: Springer.

Woodland, N. J., & Silver, B. (1949). US12241649A: United States Patent Office. Retrieved from https://worldwide.espacenet.com/patent/search/family/022402610/publication/US2612994A?q=pn%3DUS2612994

Yam, K. L., Takhistov, P. T., & Miltz, J. (2005). Intelligent packaging: concepts and applications. *Journal of Food Science*, 70(1), R1–R10.

Index

R

S

T

U

V

W

Z

Printed in the United States
by Baker & Taylor Publisher Services